RECENT STUDIES ON STRUCTURAL SAFETY AND RELIABILITY

CURRENT JAPANESE MATERIALS RESEARCH

Volume 1
Current Research on Fatigue Cracks
T. TANAKA, M. JONO and K. KOMAI

Volume 2
Statistical Research on Fatigue and Fracture
T. TANAKA, S. NISHIJIMA and M. ICHIKAWA

Volume 3
High Temperature Creep-Fatigue
R. OHTANI, M. OHNAMI and T. INOUE

Volume 4
Localized Corrosion
F. HINE, K. KOMAI and K. YAMAKAWA

RECENT STUDIES ON STRUCTURAL SAFETY AND RELIABILITY

Edited by

TAKAO NAKAGAWA

Kobe University, Japan

HIROSHI ISHIKAWA

Kagawa University, Japan

AKIRA TSURUI

Kyoto University, Japan

Current Japanese Materials Research—Vol. 5

ELSEVIER APPLIED SCIENCE
LONDON and NEW YORK

ELSEVIER SCIENCE PUBLISHERS LTD
Crown House, Linton Road, Barking, Essex IG11 8JU, England

Sole Distributor in the USA and Canada
ELSEVIER SCIENCE PUBLISHING CO., INC.
655 Avenue of the Americas, New York, NY 10010, USA

WITH 32 TABLES AND 116 ILLUSTRATIONS

British Library Cataloguing in Publication Data

Recent studies on structural safety and
reliability
1. Structures. Reliability
I. Nakagawa, Takao II. Ishikawa, Hiroshi
III. Tsurui, Akira IV. Series
6241'71

ISBN 1-85166-281-2

Library of Congress Cataloging-in-Publication Data

Recent studies on structural safety and reliability/edited by Takao Nakagawa, Hiroshi Ishikawa, Akira Tsurui.
p. cm.—(Current Japanese materials research; vol. 5)
Bibliography: p.
Includes index.
ISBN 1-85166-281-2
1. Structural stability. 2. Reliability (Engineering)
3. Structural design. I. Nakagawa, Takao, 1929–. II. Ishikawa, Hiroshi, 1941–. III. Tsurui, A. (Akira), 1942–. IV. Series.
TA656.5.R43 1989
624.1'71—dc 19 88-23522
CIP

No responsibility is assumed by the Publisher for any injury and/or damage to persons or property as a matter of products liability, negligence or otherwise, or from any use or operation of any methods, products, instructions or ideas contained in the material herein.

Printed in Northern Ireland by The Universities Press (Belfast) Ltd.

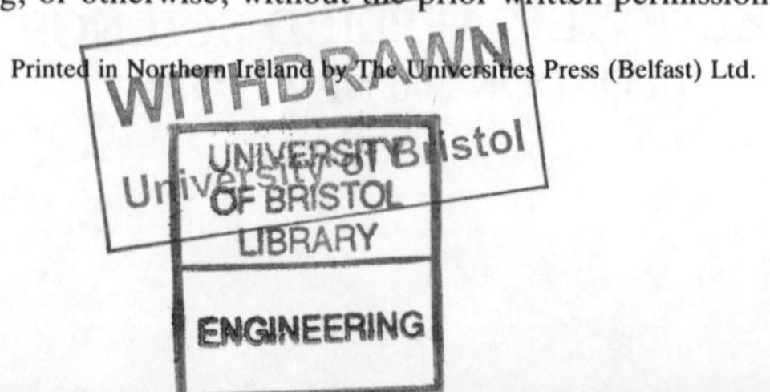

Foreword

The Current Japanese Materials Research (CJMR) series is a new publication edited by the Society of Materials Science, Japan, and published by Elsevier Science Publishers, UK, aiming at the overseas circulation of current Japanese achievement in the field of materials science and technology. This fifth volume of the series deals with *Recent Studies on Structural Safety and Reliability*. All the papers have been selected to present the most important and substantial results obtained by the authors, in order to help readers to understand the current status and recent developments of structural safety and reliability.

Although many international meetings are held every year in various specialized fields, it cannot be denied that most research results in Japan are published only in Japanese and tend, therefore, to be confined to the domestic audience. The publication of the CJMR series is an attempt to offer these results to colleagues abroad and thereby encourage the exchange of knowledge between us. I hope that our efforts will interest engineers and scientists in different countries and may contribute to the progress of materials science and technology throughout the world.

KIYOSHI OKADA
President, Society of Materials Science, Japan

Preface

Practical machines and structures are usually subjected to randomly varying external loads, and the strength of identical components will never be the same, even under the same loading conditions. In other words, both the load and strength are of an indeterminate nature. In addition, a variety of uncertainty factors will inevitably arise in the processes of their construction and maintenance.

Consequently, in order to perform rational design and maintenance, these uncertainties have to be properly evaluated on a probabilistic basis. The late Professor A. M. Freudenthal first introduced his well-known concept of failure probability to handle this problem in 1946. Following this creative research work, a number of studies have been carried out in the field where safety and reliability both play an important role.

The notion of structural safety and reliability has become of crucial importance, which is reflected by corresponding societal concern to a considerable extent. In 1969, the first International Conference on Structural Safety and Reliability (ICOSSAR '69), was held in the USA under the chairmanship of the late Professor Freudenthal. The successive second international conference (ICOSSAR '77) was held in Germany in 1977, and the third (ICOSSAR '81) in Norway in 1981. In 1985, nearly 500 participants attended the fourth conference (ICOSSAR '85), held in Japan. Furthermore, the first Japan Conference on Structural Safety and Reliability (JCOSSAR '87) was held in December 1987 under the auspices of the Japan Science Council.

Needless to say, safety and reliability play a crucial role in a variety of engineering fields such as materials science, mechanical engineering, civil and architectural engineering, naval architecture, aeronautical and space engineering, and nuclear engineering, to name but a few. Unfortunately, however, due to the shortage of space, it is impossible to deal with all the topics covering each of these engineering fields in only one volume.

In the present volume, therefore, emphasis has been placed on a limited number of topics: reliability-based design, structural reliability analysis, inspection and maintenance, and software developments for reliability assessment. In each topic, selected technical papers of vital importance are presented that can help those people concerned with structural safety and reliability to overcome difficulties, improve methodology and obtain new insight in their respective field.

The editors would like to express their utmost thanks to all the authors of the technical papers in this volume for their contributions. We also acknowledge the considerable help of Mr Yosido Fujiwara, the director of JSMS, in editing this volume.

TAKAO NAKAGAWA
HIROSHI ISHIKAWA
AKIRA TSURUI

Contents

Foreword v

Preface vii

List of Contributors xi

Reliability-Based Design

Some Problems in Probabilistic Fracture Mechanics 1
MASAHIRO ICHIKAWA and NAGATOSHI OKABE

Practical Implementation of Fatigue-Proof Design in Structural Components with the Aid of Reliability-Based Scatter Factors . 25
MASANOBU SHINOZUKA, HIROSHI ISHIKAWA and HIDETOSHI ISHIKAWA

Study on the Probabilistic Design of Fiber-Reinforced Composite Material 51
ZENICHIRO MAEKAWA

Strength of Alumina against Static and Impact Loads 71
TOSHIRO YAMADA and JUNICHI KITAZUMI

Structural Reliability Analysis

Recent Developments in the Identification of Dominant Failure Modes and Reliability Assessment for Large-Scale Frame Structures 85
YOSHISADA MUROTSU and HIROO OKADA

Reliability Analysis of Damaged Redundant Structures . . . 105
HITOSHI FURUTA, MASAAKI OHSHIMA and NARUHITO SHIRAISHI

Reliability Analysis of Bridge Structures Using a Simulation Method . . . 119
NOBUYOSHI TAKAOKA, WATARU SHIRAKI and SHIGEYUKI MATSUHO

Inspection and Maintenance

Application of the Markov Chain to the Reliability Analysis of Fatigue Crack Propagation with Non-destructive Inspection . . . 135
YOSHIHIRO SHIMADA, TAKAO NAKAGAWA and HISANOBU TOKUNO

Reliability Analysis of Fatigue Crack Growth Processes under Stationary Random Loading . . . 153
AKIRA TSURUI and AKIRA SAKO

Bayesian Reliability Analysis for Evaluating In-Service Inspection . . . 167
HIROSHI ITAGAKI, SEIICHI ITOH and NORIO YAMAMOTO

Software Developments for Reliability Assessment

Expectations from Knowledge Engineering for Reliability Improvement . . . 191
SHUICHI FUKUDA

Integrated Color-Face Graphics for Displaying Safety Conditions in an Industrial System . . . 197
FUMIO HARA

Practical Approach to the Computational Procedures for Reliability under Earthquake Conditions . . . 213
MASARU ZAKO

Index . . . 227

List of Contributors

SHUICHI FUKUDA
Welding Research Institute, Osaka University, 11-1 Mihogaoka, Ibaraki, Osaka 567, Japan

HITOSHI FURUTA
Department of Civil Engineering, Kyoto University, Yoshida-honmachi, Sakyo-ku, Kyoto, Japan

FUMIO HARA
Department of Mechanical Enginering, Science University of Tokyo, 1-3 Kagurazaka, Shinjuku-ku, Tokyo 162, Japan

MASAHIRO ICHIKAWA
Department of Mechanical and Control Engineering, University of Electro-Communications, 1-5-1 Chofugaoka, Chofu-city, Tokyo, Japan

HIROSHI ISHIKAWA
Department of Information Science, Kagawa University, 2-1 Saiwai-cho, Takamatsu City, Kagawa 760, Japan

HIDETOSHI ISHIKAWA
Department of Information Systems, Kagawa Technical College, 3202 Gunge-cho, Marugame City, Kagawa 763, Japan

HIROSHI ITAGAKI

Department of Naval Architecture and Ocean Engineering, Yokohama National University, 156 Tokiwadai, Yokohama, Kanagawa 240, Japan

SEIICHI ITOH

First Airframe Division, National Aerospace Laboratory, 6-13-1 Osawa, Mitaka, Tokyo 181, Japan

JUNICHI KITAZUMI

Department of Mechanical Engineering, Niihama National College of Technology, 7-1 Yagumocho, Niihama-city, Ehime 792, Japan

ZENICHIRO MAEKAWA

Faculty of Textile Science, Kyoto Institute of Technology, Matsugasaki, Sakyoku, Kyoto 606, Japan

SHIGEYUKI MATSUHO

Department of Civil Engineering, Tottori University, Minami 4-101, Koyama-cho, Tottori-shi 680, Japan

YOSHISADA MUROTSU

Department of Aeronautical Engineering, University of Osaka Prefecture, 4-804 Mozu-Umemachi, Sakai, Osaka 591, Japan

TAKAO NAKAGAWA

Faculty of Engineering, Kobe University, Rokkodai, Nada, Kobe 657, Japan

MASAAKI OHSHIMA

Kounoike-gumi Corporation, Osaka, Japan

NAGATOSHI OKABE

Heavy Apparatus Engineering Laboratories, Toshiba Corporation, 1-9 Suehiro-cho, Yokohama, Japan

HIROO OKADA

Department of Naval Architecture, University of Osaka Prefecture, 4-804 Mozu-Umemachi, Sakai, Osaka 591, Japan

AKIRA SAKO
Mitsubishi Heavy Industry Ltd, Hiroshima Works, Hiroshima, Japan

YOSHIHIRO SHIMADA
Olympus Optical Co., 2951 Ishikawa, Hachiooji 192, Japan

MASANOBU SHINOZUKA
Department of Civil Engineering and Operations Research, Princeton University, Engineering Quadrangle, Princeton, New Jersey 08544, USA

WATARU SHIRAKI
Department of Civil Engineering, Tottori University, Minami 4-101, Koyama-cho, Tottori-shi 680, Japan

NARUHITO SHIRAISHI
Department of Civil Engineering, Kyoto University, Yoshida-honmachi, Sakyo-ku, Kyoto, Japan

NOBUYOSHI TAKAOKA
Department of Civil Engineering, Tottori University, Minami 4-101, Koyama-cho, Tottori-shi 680, Japan

HISANOBU TOKUNO
Faculty of Engineering, Kobe University, Rokkodai, Nada, Kobe 657, Japan

AKIRA TSURUI
Department of Applied Mathematics and Physics, Kyoto University, Kyoto, Japan

TOSHIRO YAMADA
Ex-president, Niihama National College of Technology, 7-1 Yagumocho, Niihama-city, Ehime 792, Japan

NORIO YAMAMOTO
Development Department, Nippon Kaiji Kyokai, 4-7 Kioi-cho, Chiyoda-ku, Tokyo 102, Japan

MASARU ZAKO
Department of Mechanical Engineering and Technology Education, Mie University, 1515 Kamihama-cho, Tsu-city, Mie, Japan

Some Problems in Probabilistic Fracture Mechanics

MASAHIRO ICHIKAWA

Department of Mechanical and Control Engineering, University of Electro-Communications, Tokyo, Japan

and

NAGATOSHI OKABE

Heavy Apparatus Engineering Laboratories, Toshiba Corporation, Yokohama, Japan

ABSTRACT

Some of the problems in probabilistic fracture mechanics were investigated concerning the flaw size distribution and the statistical nature of the fatigue crack growth rate (da/dN). Equations for the initial flaw size distribution and for the defect detection probability during nondestructive inspection were theoretically derived. These equations were in accordance with the empirical relationships. The scatter of the fatigue crack propagation life was predicted based on the composite variability model for da/dN, this prediction being in good agreement with experimental evidence. Finally, the distribution-free approach of using the upper bound for the probability of failure is proposed. The upper bound of $P_f = \Pr(R < S)$ was first derived, and then applied to a fracture mechanics problem, where $\Pr(R < S)$ implies the probability of R being smaller than S.

INTRODUCTION

When failure of a structural component is caused by crack-like flaws, the failure condition can be expressed as $a \geq a_{cr}$, where a is the flaw size, and a_{cr} is the critical flaw size, which is a function of the fracture toughness and the applied stress. There are many factors which contribute to the uncertainty of a and a_{cr}. Among these are:

(1) (with respect to a) the statistical distribution of the initial flaw size, the defect detection probability during nondestructive

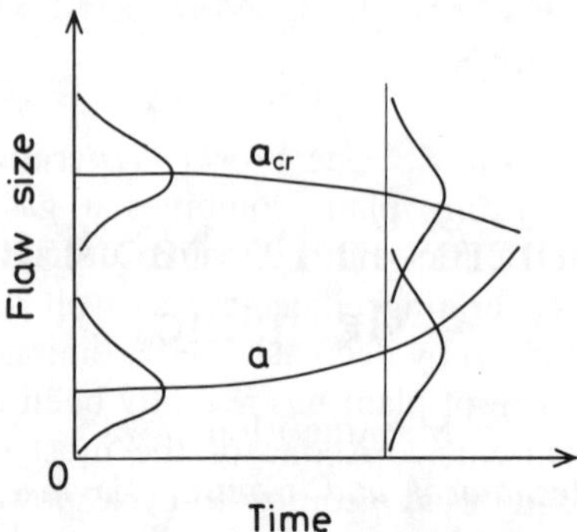

FIG. 1. Distributions of the flaw size and the critical flaw size as functions of time.

inspection, and the scatter in the initiation and propagation of cracks due to time and/or cycle dependent fracture;

(2) (with respect to a_{cr}) the scatter of fracture toughness, the scatter of external stresses and residual stresses, and the stochastic nature of corrosion.

Due to these factors, both the flaw size (a) and the critical flaw size (a_{cr}) are associated with uncertainty, and thus it is reasonable to treat a and a_{cr} not as deterministic variables but as random variables.

The distribution of a changes with time due to the initiation and propagation of cracks, the removal of flaws by in-service inspection, etc. The distribution of a_{cr} also changes with time due to corrosion, radiation damage, etc. From the distributions of a and a_{cr}, which change with time as shown in Fig. 1, the failure rate and the probability of failure can be calculated.

In order to refine the evaluation of safety and reliability based on probabilistic fracture mechanics, it is important to obtain a better understanding of the statistical nature of the various factors already mentioned. In the present paper, the flaw size distribution [1, 2] and the statistical aspect of fatigue crack growth rate [3, 4] were investigated. The distribution-free approach of using the upper bound for the probability of failure [5, 6] was also studied. Before dealing with these subjects, the necessity for probabilistic fracture mechanics in practice is discussed in the next section.

A PRACTICAL EXAMPLE: COMBINED-CYCLE POWER GENERATING PLANT

Reliability of a Combined-Cycle Power Generating Plant

This power generating plant combines a gas turbine and steam turbine system, and offers more effective utilization of coal fuel and improved power generating efficiency, as well as being effective for preventing air pollution by decreasing NO_x emissions. The commercial utilization of this type of plant has recently been increasing.

Gas turbine maintenance is one of the most significant points for assuring the reliability of the combined-cycle power generating system [7]. As shown in Fig. 2, the internal components of a gas turbine are used either continuously or intermittently in a corrosion- and erosion-attack atmosphere, and are under much more severe conditions when compared with those in a steam turbine. Therefore, outstanding creep resistance properties are required for long-term operation, and high thermal fatigue resistance properties are required for various thermal stresses due to starting and stopping. High-temperature components such as the combustion liner, transition piece, nozzle blades and moving blades might be degraded in their material properties and worn during long-term operation owing to exposure to high-temperature coal gas. Therefore, close maintenance controls, in which the exchange of worn-out components and the recovery of properties by heat-treatment are conducted, are important for high-temperature components. This is one of the significantly different points between the maintenance of gas turbines and steam turbines. If the maintenance of high-temperature components in a gas turbine is successfully performed, the gas turbine is assured of reliability as high as that of the steam turbine. Consequently, the reliability of a combined-cycle power-generating system depends on the maintenance of high-temperature components in the gas turbine.

Current Maintenance Control of High-Temperature Components in Gas Turbines

Moving Blades

Gas turbine moving blades are under quite severe conditions. They are subjected to centrifugal force and gas flow-induced torque, and are exposed to high-temperature gas. To estimate closely the degradation in material properties of the moving blades is one of the most

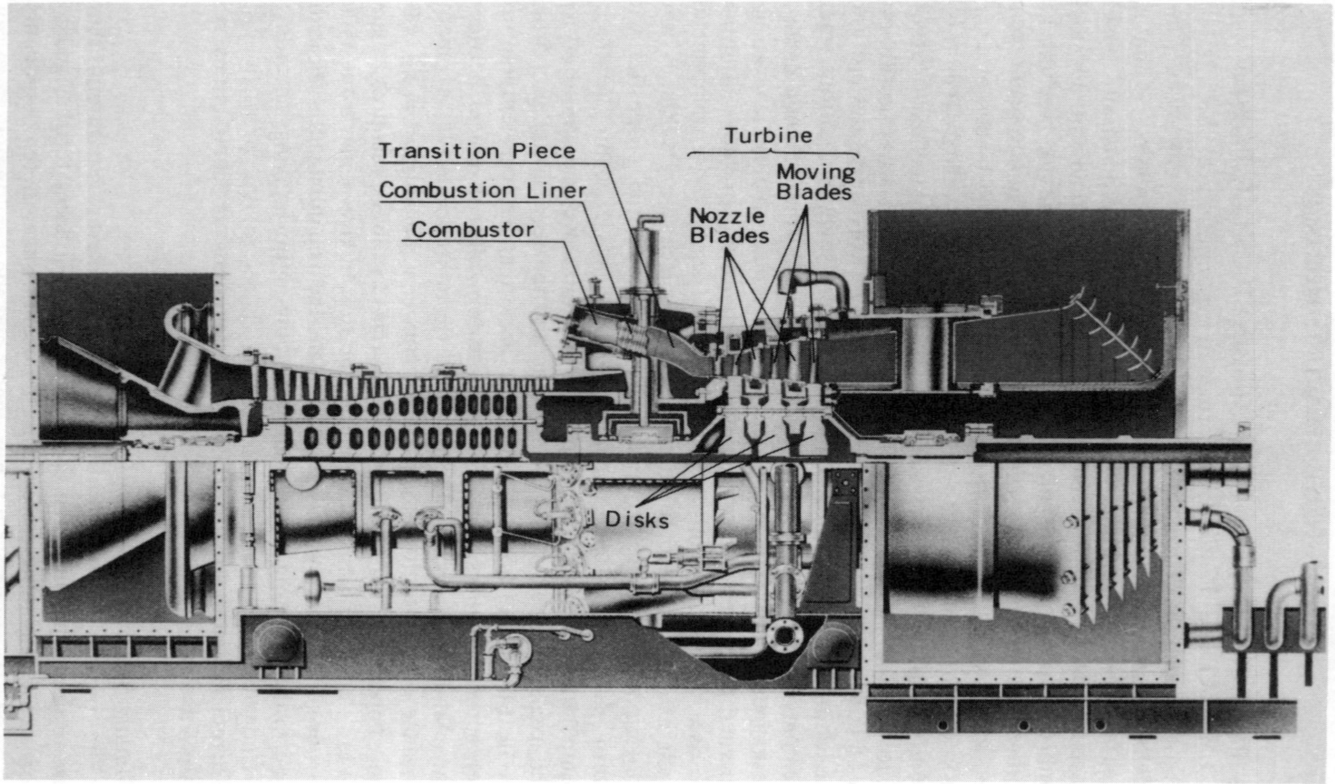

Fig. 2. Gas turbine components.

significant aspects for periodical checking in service. Checks of the moving blades are generally performed at every periodical check, after removing each blade from its dovetail location in the rotor disks. The oxidized scale on their surfaces is removed by liquid honing, so that a nondestructive inspection (visual detection, dye-penetrant test, fluorescent-penetrant test, etc.) can be performed on them. On the other hand, material property investigations and tests of mechanical strength properties (creep–rupture, hardness, tensile and impact tests) are conducted on pieces cut from one or two moving blades of each disk. The procedures to be performed at the next periodical check (i.e. whether the moving blades can be used continuously in their existing state, or need to be heat-treated to recover the material properties, or need to be exchanged with new blades) are usually determined by estimating the material degradation from the preceding test results. In moving blades, crack initiation is unacceptable. Hence, when cracks are detected, the moving blades must be exchanged with new ones. However, damage-tolerant design will challenge this problem in future for both the moving blades and disks [8, 9].

Nozzle blades

The nozzle blades are also exposed to high-temperature coal gas flows. However, as the nozzle blades are stationary and subjected to rather lower stresses than the moving blades, crack initiation in the nozzle blades as shown in Fig. 3 is mainly due to thermal stress fatigue, so that the situation is different from that of moving blades. Thus, when periodically checking in service, the nozzle blades of gas turbines are generally detached from between the retaining ring and the supporting ring. After removing the oxidized scale on their surfaces by liquid honing, a defect inspection by dye-penetrant tests is performed. After that, those nozzle blades with only slight cracks are used in their existing state or after grinding and overlaying, according to judgement on the detected cracks. Therefore, the prediction of crack propagation life up to the next periodical check is critical to assure the reliability of nozzle blades.

Combustion Liner and Transition Piece

The combustion liner and transition piece are constructed from thin plates and, being subjected to very low stresses, no material degradation due to applied stresses occurs. However, since these components have complex forms as shown in Fig. 4 and the local temperature can

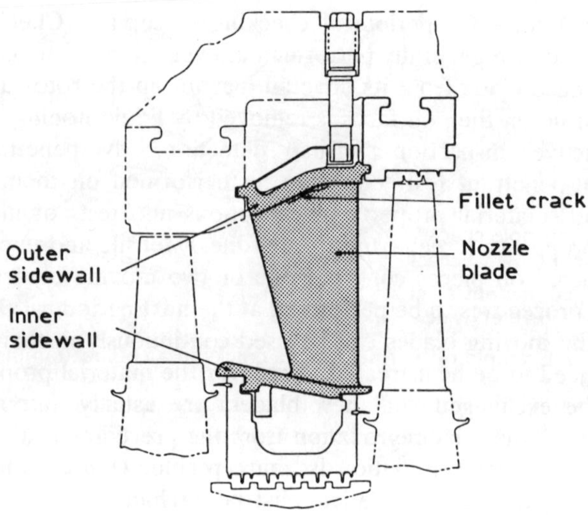

FIG. 3. Example of cracking initiated in a nozzle blade.

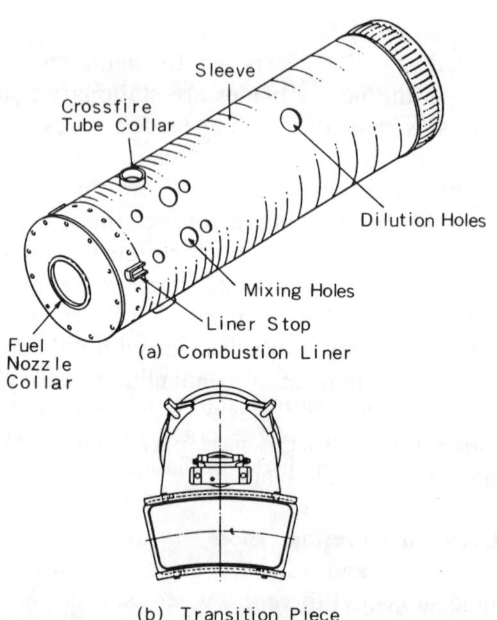

FIG. 4. Combustion liner and transition piece.

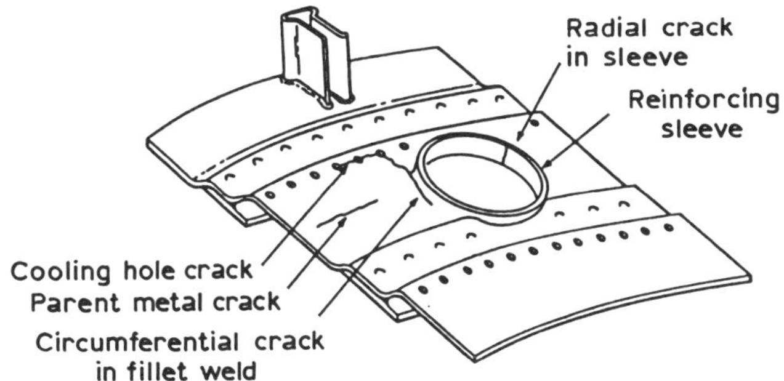

FIG. 5. Example of thermal fatigue cracks in a combustion liner.

exceed 1000 °C in spite of air-cooling of the internal surface, some burn damage or hair-line cracks, as shown in Fig. 5, often occur. When such defects are detected during periodical checks in service, the maintenance for hair-line crack initiation usually involves welding repairs if they cannot be used in the existing state. Only in the case of burn damage are they replaced by new components. Thus, the prediction of crack propagation life up to the next periodical check has significant effects on the reliability of these components.

Predicting the Crack Propagation Life of Gas Turbine Components in Periodical Maintenance Checks

As already mentioned, when cracks are detected in the nozzle blades, combustion liner and transition piece, the prediction of crack propagation life is essential for judging whether these components can be used without welding repair or need to be replaced with new ones. In commercial use applications, the prediction of crack propagation life at each periodical check is significant for assuring the reliability of gas turbines. In this prediction, the application of probabilistic fracture mechanics is recommended to take account of the uncertainty of relevant factors, and investigations into the flaw size distribution and the statistical aspects of the fatigue crack propagation rate (da/dN) are important for such probabilistic predictions.

FLAW SIZE DISTRIBUTION [1, 2]

Significance of Flaw Size Distribution

The flaw size distribution is one of the most important factors in reliability analysis. It has been shown that the calculated probability of failure is quite sensitive to the flaw size distribution function that has been assumed [10]. When nondestructive inspection (NDI) is not performed before putting into service, the initial flaw size distribution is relevant to reliability analysis. When NDI is performed and the detected flaws are repaired, the after-NDI flaw size distribution is then relevant. In estimating the after-NDI flaw size distribution, the defect detection probability plays an important role. Thus, in this section we are concerned with the initial flaw size distribution and the defect detection probability.

Up to the present time, a variety of distribution functions have been employed by different authors for the flaw size. These include exponential, Weibull, normal, log-normal, gamma, beta, triangular and Jonson–Su distributions. Similarly, a variety of functional forms have been used for the defect detection probability (P_D). These include $P_D = 1 - \exp[-A(a - a_1)]$, $P_D = 1 - \exp[-B(a - a_1)^\alpha]$, $P_D = 1 - (1/2)\,\mathrm{erfc}[\nu \ln(a/a_1)]$, $P_D = B(a - a_1)^\alpha$ and $P_D = [1 - (a_1/a)]^\delta$. In order to resolve this confusion, theoretical investigations into the underlying nature of the initial flaw size distribution and the defect detection probability seem useful, in addition to the accumulation of experimental data. From this point of view, some theoretical aspects are considered next.

Initial Flaw Size Distribution [1]

Suppose for a population of specimens that the following assumptions are made:

(A1) a large number of flaws are involved in each specimen, and the weakest link model is applicable;
(A2) fracture is governed by the principle of linear elastic fracture mechanics;
(A3) the fracture strength of specimens follows a two-parameter Weibull distribution.

These assumptions seem to be reasonable in a wide range of applications including the fracture of engineering ceramics. Let $F(\sigma_f)$, $G(a)$ and $G_L(a_{max})$ be the distribution functions of the fracture strength (σ_f) of a specimen, a the individual flaw size, and a_{max} the maximum flaw size in a specimen. The third assumption implies

$$F(\sigma_f) = 1 - \exp\left[-\left(\frac{\sigma_f}{\sigma_0}\right)^m\right] \tag{1}$$

$$m > 0, \qquad \sigma_0 > 0$$

where m and σ_0 are the shape and scale parameters, respectively. From the first and second assumptions, σ_f is related to a_{max} as

$$\sigma_f = \frac{K_0}{\sqrt{a_{max}}} \tag{2}$$

where K_0 depends on the crack geometry. For example, when circular internal flaws in a sufficiently large component are considered, $K_0 = \sqrt{\pi}\, K_{Ic}/2$, where K_{Ic} is the plane strain fracture toughness. From Eqns. (1) and (2), the distribution function of the maximum flaw size $G_L(a_{max})$ is obtained as follows:

$$G_L(a_{max}) = \exp\left[-\left(\frac{a_{max}}{K_0^2/\sigma_0^2}\right)^{-m/2} \right] \tag{3}$$

Equation (3) has the form of a second asymptotic distribution of the largest value with a location parameter of zero.

Equation (3) gives the distribution of the size of the largest flaw among a large number of flaws in a specimen. According to the statistics of extremes, the second asymptotic distribution given by Eqn. (3) is obtained when the individual flaw size (a) has the following distribution function $G(a)$ in the range for large values of a:

$$G(a) = 1 - \left\lfloor \frac{a}{L} \right\rfloor^{-\lambda} \quad \text{(for large } a\text{)} \tag{4}$$

$$\lambda > 0, \qquad L > 0$$

where L has the dimension of length. When $G(a)$ is given by Eqn. (4) in the range for large a, the probability density function of the individual flaw size $g(a)$ has the following form in the range for large a:

$$g(a) = \frac{\lambda}{L}\left[\frac{a}{L}\right]^{-(1+\lambda)} \quad \text{(for large } a\text{)}$$

$$\lambda > 0, \qquad L > 0 \tag{5}$$

It is to be noted that Eqn. (5) is of the same form as the empirical relationship $pa^{\delta} = \text{const.}$ found by Nakamura [11] from several statistical sets of data of flaw sizes, where p is the frequency of appearance.

The result of $G(a) = 1 - (a/L)^{-\lambda}$ does not hold in the range for small values of a. However, this is not a serious drawback of Eqn. (4) since small flaws are far less important than large flaws from the viewpoint of reliability analysis. In other words, it is much more important to know the flaw size distribution in the range for large a than in the range for small a. One of the distribution functions which has the properties of Eqn. (4) is

$$G(a) = \exp\left[-\left(\frac{a}{L}\right)^{-\lambda}\right] \quad (0 \le a < \infty) \tag{6}$$

which itself has the form of a second asymptotic distribution of the largest value. Equation (6) does not have a theoretical basis in the range for small a. However, when an accurate description of $G(a)$ in the range for small a is not important, Eqn. (6) might be used for the convenience of a simple analytical expression.

Defect Detection Probability [2]

Next, we consider the defect detection probability (P_D) of non-destructive inspection. P_D is a function of the flaw size (a) and also depends on such NDI methods as ultrasonic or X-ray inspection. We consider here the flaw size dependence of P_D that can be expected to apply to the various NDI methods.

Suppose that a population of specimens is subjected to NDI, and that the flaws detected are perfectly repaired. The following assumptions are made:

(B1) a large number of flaws exist in each specimen after repair as well as before repair;
(B2) fracture before and after repair is governed by the principle of linear elastic fracture mechanics;
(B3) fracture strength of specimens before repair and that after repair follow two-parameter Weibull distributions.

These are simply extensions of the previous assumptions (A1)–(A3) to the situation after repair.

In a way similar to the derivation of Eqn. (3), the distribution function of the maximum flaw size after repair, $G_L^*(a_{max})$, can be obtained from the assumptions (B1)–(B3) as

$$G_L^*(a_{max}) = \exp\left[-\left(\frac{a_{max}}{K_0^2/\sigma_0^{*2}}\right)^{-m^*/2}\right] \tag{7}$$

where m^* and σ_0^* are the shape and scale parameters, respectively, of the two-parameter Weibull distribution of the fracture strength after repair. K_0 is the same factor as that in Eqn. (2). Thus, the distribution function of the individual flaw size after repair, $G^*(a)$, can be expressed in the following form:

$$G^*(a) = 1 - \left(\frac{a}{L^*}\right)^{-\lambda^*} \quad \text{(for large } a\text{)} \tag{8}$$

$$\lambda^* > 0, \qquad L^* > 0$$

The probability density function, $g^*(a)$, corresponding to $G^*(a)$ is

$$g^*(a) = \frac{\lambda^*}{L^*}\left[\frac{a}{L^*}\right]^{-(1+\lambda^*)} \quad \text{(for large } a\text{)} \tag{9}$$

Let n and n^* be the number of flaws in a specimen before and after repair, respectively. The defect detection probability (P_D) is obtained for large values of a as

$$\begin{aligned} P_D &= 1 - \frac{n^* g^*(a)\,\mathrm{d}a}{n g(a)\,\mathrm{d}a} \\ &= 1 - \frac{n^*(\lambda^*/L^*)(a/L^*)^{-(1+\lambda^*)}}{n(\lambda/L)(a/L)^{-(1+\lambda)}} \quad \text{(for large } a\text{)} \end{aligned} \tag{10}$$

which can be rewritten in the following form:

$$P_D = 1 - \left[\frac{a}{L_1}\right]^{\lambda-\lambda^*} \quad \text{(for large } a\text{)} \tag{11}$$

where L_1 has the dimension of length. Noting that P_D should increase with an increase in a, it follows that $\lambda - \lambda^*$ should be negative. Hence

$$P_D = 1 - \left[\frac{a}{L_1}\right]^{-\kappa} \quad \text{(for large } a\text{)} \tag{12}$$

$$\kappa > 0, \qquad L_1 > 0$$

It is interesting to note that Eqn. (12) has the same form as the empirical relation $(1 - P_D)a^{\gamma} = \text{const.}$ found by Nakamura [11] for a magnetic particle inspection of welds. The Nakatsuji *et al.* equation [12] $P_D = (1 - a_1/a)^{\nu}$ $(a \geq a_1)$ that was found for an ultrasonic inspection of welds also reduces to the form of Eqn. (12) in the range for large a.

STATISTICAL ASPECTS OF FATIGUE CRACK GROWTH RATE [3, 4]

Composite Variability Model for Parameter Randomization

In this section, we present our recent progress with the statistical aspects of fatigue crack growth rate which was attained after our report [13] in Vol. 2 of this series.

In order to incorporate the scatter of the fatigue crack growth rate ($\mathrm{d}a/\mathrm{d}N$) into reliability analysis, it is convenient to randomize C and/or m in Paris's law

$$\frac{\mathrm{d}a}{\mathrm{d}N} = C(\Delta K)^m \tag{13}$$

In Ref. 13, we examined two models for this randomization, the inter-specimen variability and the intra-specimen variability models. In the former model, C and m were taken to vary between specimens, but not to vary in space within a specimen. In the latter model, C was taken to vary in space within a specimen. The statistical properties of the crack propagation life predicted by each model was in reasonable but insufficient agreement with the experimental results. The subsequent efforts to find a more accurate prediction based on the composite variability model is described in this section.

The composite variability model is a combination of the inter-specimen and intra-specimen variability models [4]. In the composite variability model, C is taken to be a random variable whose randomness is composed of the variability within a specimen and that between specimens. The value of m is taken to be constant. With respect to the variability of C within a specimen, we employ the discrete approximation [14] that was applied in the analysis [13] of the intra-specimen variability model. In this approximation, the spatial variation of C within a specimen is treated as occurring at intervals of finite distance δ as shown in Fig. 6. It should be noted that this discrete approximation takes account of the correlation between values of C at two different points in space. The values of C at any two points are completely correlated in the extreme case of $\delta \to \infty$, and completely uncorrelated in the other extreme case of $\delta \to 0$. In between these two cases, the degree of correlation increases with an increase in δ. The term δ is the same as the correlation distance defined by Ishikawa and Tsurui [15].

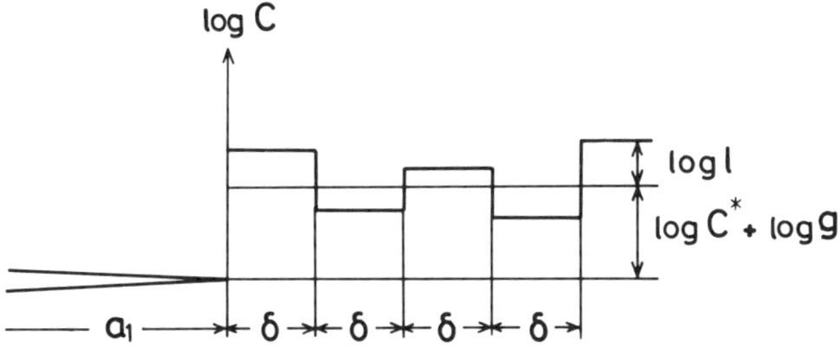

FIG. 6. Composite variability model.

Prediction by the Composite Variability Model

Let us calculate the crack propagation life based on the composite variability model. Using the discrete approximation just mentioned, we can express the parameter C as

$$C = C^* g l \tag{14}$$

or

$$\log C = \log C^* + \log g + \log l \tag{15}$$

where g and l are random variables representing the inter-specimen variability and the intra-specimen variability of C, respectively. We assume that both $\log g$ and $\log l$ follow normal distributions with zero mean. That is,

$$\begin{aligned} \mu(\log g) &= 0 \\ \mu(\log l) &= 0 \end{aligned} \tag{16}$$

For comparison in our experiment, the crack propagation life is calculated for the center-cracked specimen subjected to a constant stress range ($\Delta\sigma$) shown in Fig. 7. Suppose n intervals of length δ are

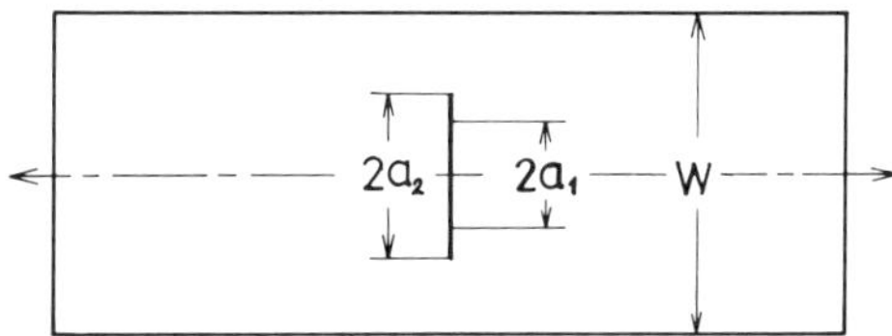

FIG. 7. Center cracked specimen.

involved between the initial crack length $a = a_1$ and a certain length $a = a_2$. Let $C^* gl_i$ be the value of C in the ith interval $a_1 + (i-1)\delta < a < a_1 + i\delta$, and N_i be the number of cycles required to pass through the ith interval. From Eqn. (13), we can obtain

$$N_i = \frac{1}{C^*(\Delta\sigma\sqrt{\pi})^m} \cdot \frac{1}{gl_i} \int_{a_1+(i-1)\delta}^{a_1+i\delta} \left(a \sec \frac{\pi a}{W}\right)^{-m/2} \mathrm{d}a \tag{17}$$

Here, the following expression was used for ΔK of the center-cracked specimen

$$\Delta K = \Delta\sigma\sqrt{\pi a \sec(\pi a/W)} \tag{18}$$

The crack propagation life N for crack growth from $a = a_1$ to $a = a_2$ can be obtained from Eqn. (17) as

$$N = \sum N_i = \frac{1}{C^*(\Delta\sigma\sqrt{\pi})^m} \frac{1}{g} \sum_{i=1}^{n} \frac{I_i}{l_i} \tag{19}$$

where

$$I_i = \int_{a_1+(i-1)\delta}^{a_1+i\delta} \left(a \sec \frac{\pi a}{W}\right)^{-m/2} \mathrm{d}a \tag{20}$$

The mean value of N can be obtained from Eqn. (20) as

$$\mu_N = \frac{1}{C^*(\Delta\sigma\sqrt{\pi})^m} \mu\left(\frac{1}{g}\right) \mu\left(\frac{1}{l}\right) \sum_{i=1}^{n} I_i \tag{21}$$

where the mean value of the variable x is denoted as either μ_x or $\mu(x)$. From the assumption that both g and l follow a log-normal distribution, it follows that

$$\mu\left[\frac{1}{g}\right] = \frac{1+\eta_g^2}{\mu_g}, \qquad \mu\left[\frac{1}{l}\right] = \frac{1+\eta_l^2}{\mu_l} \tag{22}$$

where η_g and η_l are the coefficient of variation of g and l, respectively. From Eqns. (21) and (22), we can obtain

$$\mu_N = \frac{1}{C^*(\Delta\sigma\sqrt{\pi})^m} \frac{1+\eta_g^2}{\mu_g} \frac{1+\eta_l^2}{\mu_l} \int_{a_1}^{a_2} \left(a \sec \frac{\pi a}{W}\right)^{-m/2} \mathrm{d}a \tag{23}$$

The variance of N can be obtained from Eqn. (19) as

$$\sigma_N^2 = \left[\frac{1}{C^*(\Delta\sigma\sqrt{\pi})^m}\right]^2 \left[\sigma^2\left(\frac{1}{g}\right)\sigma^2\left(\sum_{i=1}^{n}\frac{I_i}{l_i}\right) + \left\{\mu\left(\frac{1}{g}\right)\right\}^2 \sigma^2\left(\sum\frac{I_i}{l_i}\right) + \left\{\mu\left(\sum\frac{I_i}{l_i}\right)\right\}^2 \sigma^2\left(\frac{1}{g}\right)\right] \tag{24}$$

where the variance of the variable x is denoted as either σ_x^2 or $\sigma^2(x)$. $\sigma^2(\sum (I_i/l_i))$ can be approximated as

$$\sigma^2\left[\sum\frac{I_i}{l_i}\right] = \sigma^2\left[\frac{1}{l}\right]\sum I_i^2 \simeq \sigma^2\left[\frac{1}{l}\right]\delta\int_{a_1}^{a_2}\left(a\sec\frac{\pi a}{W}\right)^{-m} \mathrm{d}a \tag{25}$$

The approximation in Eqn. (25) is fairly good when δ/a is small. Since g and l follow log-normal distributions, the following relationships hold:

$$\begin{aligned}\sigma^2\left[\frac{1}{g}\right] &= \left[\frac{1+\eta_g^2}{\mu_g}\right]^2 \eta_g^2 \\ \sigma^2\left[\frac{1}{l}\right] &= \left[\frac{1+\eta_l^2}{\mu_l}\right]^2 \eta_l^2\end{aligned} \tag{26}$$

Using Eqns. (25) and (26), Eqn. (24) becomes

$$\sigma_N^2 \simeq \left[\frac{1}{C^*(\Delta\sigma\sqrt{\pi})^m}\cdot\frac{(1+\eta_g^2)(1+\eta_l^2)}{\mu_g\mu_l}\right]^2 \times\left[\eta_l^2(1+\eta_g^2)\delta\int_{a_1}^{a_2}\left(a\sec\frac{\pi a}{W}\right)^{-m}\mathrm{d}a + \eta_g^2\left[\int_{a_1}^{a_2}\left(a\sec\frac{\pi a}{W}\right)^{-m/2}\mathrm{d}a\right]^2\right] \tag{27}$$

When the variable x follows a log-normal distribution, the relationship $\mu(\ln x) = \ln(\mu_x/\sqrt{1+\eta_x^2})$ holds. Using this relationship and Eqn. (16), it follows that

$$\begin{aligned}\mu_g &= \sqrt{1+\eta_g^2} \\ \mu_l &= \sqrt{1+\eta_l^2}\end{aligned} \tag{28}$$

Substituting Eqn. (28) into Eqns. (23) and (27), μ_N and σ_N^2 can be expressed as

$$\mu_N = \frac{\sqrt{(1+\eta_g^2)(1+\eta_l^2)}}{C^*(\Delta\sigma\sqrt{\pi})^m}\int_{a_1}^{a_2}\left(a\sec\frac{\pi a}{W}\right)^{-m/2} \mathrm{d}a \tag{29}$$

$$\begin{aligned}\sigma_N^2 \simeq &\left[\frac{1}{C^*(\Delta\sigma\sqrt{\pi})^m}\right]^2(1+\eta_g^2)(1+\eta_l^2)\\ &\times\left[\eta_l^2(1+\eta_g^2)\delta\int_{a_1}^{a_2}\left(a\sec\frac{\pi a}{W}\right)^{-m}\mathrm{d}a\right.\\ &\left.+\,\eta_g^2\left[\int_{a_1}^{a_2}\left(a\sec\frac{\pi a}{W}\right)^{-m/2}\mathrm{d}a\right]^2\right]\end{aligned} \tag{30}$$

The coefficient of variation of the crack propagation life, η_N, can be obtained from Eqns. (29) and (30) as

$$\eta_N \simeq \left[\eta_g^2 + \delta\eta_l^2(1+\eta_g^2)\frac{\displaystyle\int_{a_1}^{a_2}[a\sec(\pi a/W)]^{-m}\,\mathrm{d}a}{\left\{\displaystyle\int_{a_1}^{a_2}[a\sec(\pi a/W)]^{-m/2}\,\mathrm{d}a\right\}^2}\right]^{1/2} \tag{31}$$

In the foregoing derivation, g and l have been assumed to follow log-normal distributions. When this assumption is not made, μ_N, σ_N^2 and η_N can be derived by using a first-order approximation. The results for such a case are essentially the same as those here, except that the factors $1+\eta_g^2$ and $1+\eta_l^2$ in Eqns. (29)–(31) are replaced by unity.

Comparison with Experimental Results

The theoretical prediction of η_N (Eqn. (31)) by the composite variability model will now be compared with our experimental results. The details of this experiment were reported in Ref. 13, and only a brief description is given here. The material used was 2024-T3 aluminum alloy plate of 1 mm thickness. The center-crack specimen with a width $W = 70$ mm and half-crack length of 7 mm was machined, and tested under conditions of the stress range $\Delta\sigma = 59{\cdot}4$ MPa and the stress ratio $R = 0{\cdot}2$. Thirty specimens were tested under identical conditions. The power law $\mathrm{d}a/\mathrm{d}N = C_s(\Delta K)^{m_s}$ was applied to the result of each specimen. It should be noted that C_s and m_s are different from C and m in Eqn. (13). That is, Eqn. (13) was applied to each point in space, whereas $\mathrm{d}a/\mathrm{d}N = C_s(\Delta K)^{m_s}$ was applied to the specimen as a whole. It can be shown theoretically that the expected values of m_s and $\log C_s$ coincide with m and $\log C^*$, respectively. The sample means of m_s and $\log C_s$ were 2·939 and −7·156, respectively. Hence, in the following comparison, values of $m = 2{\cdot}939$ and $\log C^* = -7{\cdot}156$ are assumed.

Let us compare Eqn. (31) with the experimental results. The initial crack length (a_1) was taken as $a_1 = 9$ mm. The open circles in Fig. 8 show the experimental values of η_N as a function of the final crack length (a_2). The values of η_g and $\eta_l\sqrt{\delta/a_1}$ in Eqn. (31) were determined so that the sum of the squares of the difference between the theoretical and experimental values was minimized. The results were $\eta_g = 0{\cdot}103$ and $\eta_l\sqrt{\delta/a_1} = 0{\cdot}0534$. Using these values, Eqn. (31) is shown as the solid curve in Fig. 8. The dotted curves show the

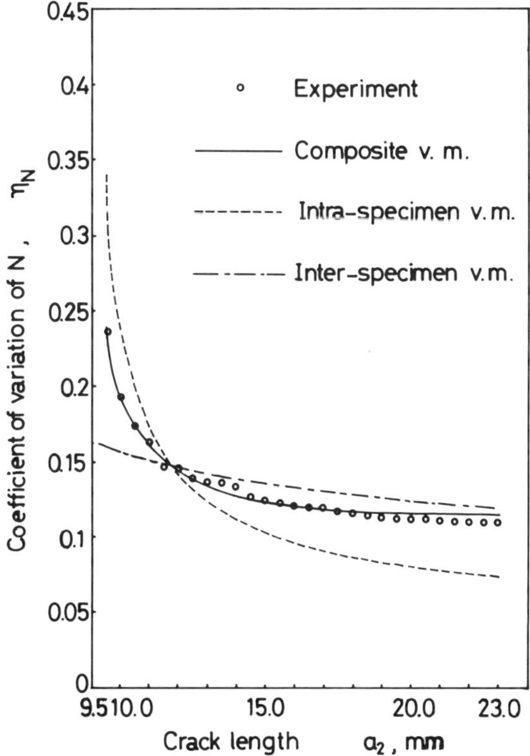

FIG. 8. Comparison between theory and experiment (v.m. denotes the variability model).

theoretical prediction based on the inter-specimen variability model and that based on the intra-specimen variability model. It can be seen that the composite variability model is in better agreement than are the other two models.

APPROACH BASED ON THE UPPER BOUND FOR PROBABILITY OF FAILURE

Derivation using the Mean and Variance [5,6]

Suppose that fracture is governed by the condition of $K_{\mathrm{I}} > K_{\mathrm{Ic}}$ or $\alpha\sigma\sqrt{\pi a} > K_{\mathrm{Ic}}$, where K_{I} is the mode I stress intensity factor, K_{Ic} is the plane strain fracture toughness, σ is the applied stress, a is the crack length, and α is a geometrical factor. When the distribution forms of σ, a and K_{Ic} are unclear and only their means and variances are known, the second moment method using the reliability index (or safety index) β is useful. In this section, we propose another second moment approach using the upper bound for the probability of failure (P_{f}).

With this approach, the upper bound of P_{f} is evaluated from the means and variances of the relevant variables, and is used as a relative measure for comparing the reliability or safety of two similar structures. When this upper bound is used as an absolute measure for the probability of failure, it naturally gives a conservative result.

Before considering the fracture mechanics problem, we will first consider the upper bound of $P_{\mathrm{f}} = \Pr(R < S)$, where R and S are the strength and the applied stress measurements, respectively, and Pr( ) denotes the probability of occurrence of the event in parenthesis. The following equation was derived by Thien and Massoud [16] as such an upper bound:

$$P_{\mathrm{f}} = \frac{\nu_{\mathrm{c}}^2[(1+\eta_R^2)(1+3\eta_S^2) + k^*(k^*-2)(1+\eta_S^2)^2]}{[k^*\nu_{\mathrm{c}}(1+\eta_S^2)-1]^2}$$
$$k^* = \frac{\nu_{\mathrm{c}}[(1+\eta_R^2)(1+3\eta_S^2)/(1+\eta_S^2)]-1}{\nu_{\mathrm{c}}(1+\eta_S^2)-1} \qquad (32)$$

where $\nu_{\mathrm{c}}(=\mu_R/\mu_S)$ is the central factor of safety. μ_R, σ_R^2 and η_R, and μ_S, σ_S^2 and η_S are the mean, variance and coefficient of variation of R and S, respectively. In the following analysis, we will derive a more powerful equation which can yield a lower upper bound than Eqn. (32).

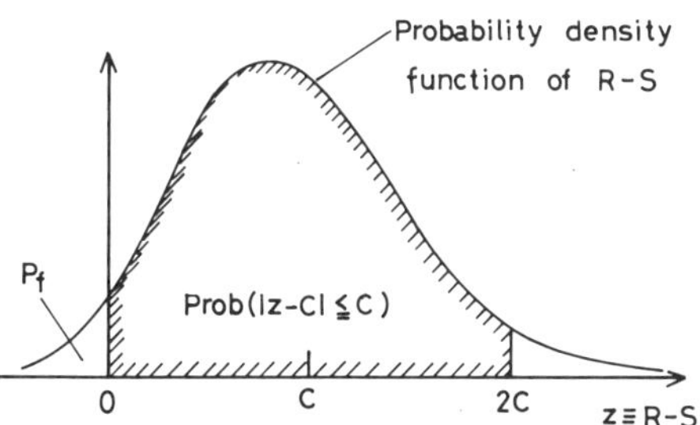

FIG. 9. Distribution of $R - S$.

Let $z = R - S$. Referring to Fig. 9, it follows that

$$P_f = \Pr(z < 0) < \Pr(|z - c| \geq c) \tag{33}$$

where c is an arbitrary positive value. From the Bienayme–Chebyshev inequality, we can obtain

$$\Pr(|z - c| \geq c) \leq \frac{E[(z - c)^2]}{c^2} \tag{34}$$

Noting that $E[(z - c)^2] = E[z^2] - 2c\mu_z + c^2$ and $E[z^2] = \mu_z^2 + \sigma_z^2$, it can be shown that $E[(z - c)^2]/c^2$ takes the minimum of $\sigma_z^2/(\mu_z^2 + \sigma_z^2)$ at $c = (\mu_z^2 + \sigma_z^2)/\mu_z$, where μ_z and σ_z^2 are the mean and variance of z, respectively. Thus, we can obtain

$$P_f < \frac{\sigma_z^2}{\mu_z^2 + \sigma_z^2} \tag{35}$$

When R and S are independent of each other, Eqn. (35) becomes

$$P_f < \frac{\sigma_R^2 + \sigma_S^2}{(\mu_R - \mu_S)^2 + \sigma_R^2 + \sigma_S^2} \tag{36}$$

In terms of the central factor of safety (ν_c), Eqn. (36) can be written as

$$P_f < \frac{\nu_c^2\eta_R^2 + \eta_S^2}{(\nu_c - 1)^2 + \nu_c^2\eta_R^2 + \eta_S^2} \tag{37}$$

Equation (37) is compared with Eqn. (32) for the case of $\eta_R = 0{\cdot}05$ and $\eta_S = 0{\cdot}1$ in Fig. 10. It can be clearly seen that Eqn. (37) is more powerful than Eqn. (32). Furthermore, Eqn. (32) is not exact in that its derivation required the use of a first-order approximation, whereas no approximation was used in deriving Eqn. (37).

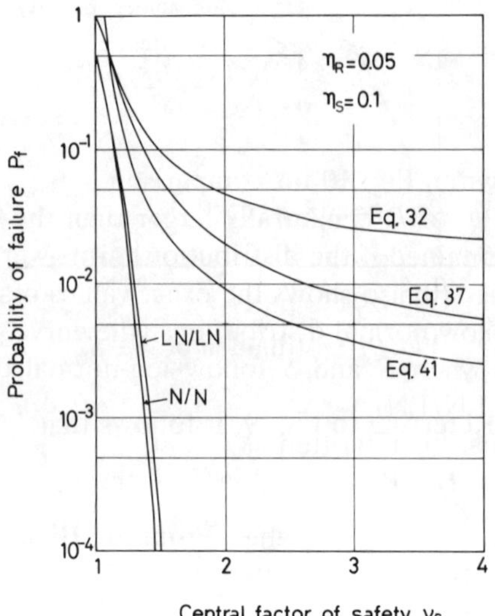

FIG. 10. Upper bound of $P_f = \Pr(R < S)$.

Equation (37) can be applied to any case. If the conditions that the distribution of $R-S$ is continuous and unimodal, and the mode of $R-S$ coincides with its mean are imposed, an equation even more powerful than Eqn. (37) can be obtained. Under these conditions, the Camp–Meidell inequality gives

$$\Pr(|z-\mu_z| \geq \mu_z) \leq \frac{4}{9} \cdot \frac{\sigma_z^2}{\mu_z^2} \tag{38}$$

Hence, it follows that

$$P_f < \frac{4}{9} \cdot \frac{\sigma_z^2}{\mu_z^2} \tag{39}$$

When R and S are independent of each other, Eqn. (39) can be written as

$$P_f < \frac{4}{9} \cdot \frac{\sigma_R^2 + \sigma_S^2}{(\mu_R - \mu_S)^2} \tag{40}$$

In terms of the central factor of safety (ν_c), Eqn. (40) can be rewritten as

$$P_{\mathrm{f}} < \frac{4}{9} \cdot \frac{\nu_{\mathrm{c}}^2 \eta_R^2 + \eta_S^2}{(\nu_{\mathrm{c}} - 1)^2} \tag{41}$$

which is also shown in Fig. 10 for comparison.

The upper bound of P_{f} is naturally larger than the exact value of P_{f} which would be obtained if the distribution forms were known. To see this difference, Fig. 10 also shows the exact values of P_{f} for the case in which R and S follow normal distributions (the curve denoted as N/N) and for the case when R and S follow log-normal distributions (the curve denoted as LN/LN).

Equation (37) can be rewritten as

$$\nu_{\mathrm{c}} < \frac{P_{\mathrm{f}} + [P_{\mathrm{f}}^2 - \{(1 + \eta_R^2)P_{\mathrm{f}} - \eta_R^2\}\{(1 + \eta_S^2)P_{\mathrm{f}} - \eta_S^2\}]^{1/2}}{(1 + \eta_R^2)P_{\mathrm{f}} - \eta_R^2} \tag{42}$$

Similarly, Eqn. (41) can be rewritten as

$$\nu_{\mathrm{c}} < \frac{2{\cdot}25P_{\mathrm{f}} + [2{\cdot}25P_{\mathrm{f}}(\eta_R^2 + \eta_S^2) - \eta_R^2\eta_S^2]^{1/2}}{2{\cdot}25P_{\mathrm{f}} - \eta_R^2} \tag{43}$$

Equations (42) and (43) give the upper bound for the central factor of safety (ν_{c}) which is needed to make the probability of failure lower than a specified value. These equations were used in the reliability evaluation system for ceramic gas turbines by Hamanaka *et al.* [17].

Equation (41) was obtained under the conditions that the distribution of $R - S$ was continuous and unimodal, and that the mode of $R - S$ coincided with its mean. A recent investigation [18] suggests that Eqn. (41) can be used even when the mode of $R - S$ deviates from its mean.

Application to Unstable Fracture

Consider unstable fracture in a linear elastic fracture mechanics regime. The limit state equation can then be expressed as $K_{\mathrm{Ic}} - \alpha\sigma\sqrt{\pi a} = 0$. In order to avoid the first-order approximation, we choose R and S as

$$\begin{aligned} R &= \ln K_{\mathrm{Ic}} \\ S &= \ln \alpha\sigma\sqrt{\pi a} \end{aligned} \tag{44}$$

When α is regarded as constant, we can obtain

$$\begin{aligned} \mu_R &= \mu[\ln K_{\mathrm{Ic}}] \\ \mu_S &= \ln(\alpha\sqrt{\pi}) + \mu[\ln \sigma] + \tfrac{1}{2}\mu[\ln a] \\ \sigma_R^2 &= \sigma^2[\ln K_{\mathrm{Ic}}] \\ \sigma_S^2 &= \sigma^2[\ln \sigma] + \tfrac{1}{4}\sigma^2[\ln a] \end{aligned} \tag{45}$$

where $\mu(\cdot)$ and $\sigma^2(\cdot)$ denote the mean and variance, respectively. Substitution of Eqn. (45) into Eqn. (37) or Eqn. (41) gives the upper bound of $P_{\mathrm{f}} = \Pr(K_{\mathrm{Ic}} < \alpha\sigma\sqrt{\pi a})$.

When the means and variances of $\ln K_{\mathrm{Ic}}$, $\ln \sigma$ and $\ln a$ are unknown, but the means and variances of K_{Ic}, σ and a are known, it is necessary to use the first-order approximation. Using this approximation, Eqn. (45) becomes

$$\begin{aligned} \mu_R &\simeq \ln \mu_{K_{\mathrm{Ic}}} \\ \mu_S &\simeq \ln(\alpha\mu_\sigma\sqrt{\pi\mu_a}) \\ \sigma_R^2 &\simeq \eta_{K_{\mathrm{Ic}}}^2 \\ \sigma_S^2 &\simeq \eta_\sigma^2 + \tfrac{1}{4}\eta_a^2 \end{aligned} \tag{46}$$

where $\eta_{K_{\mathrm{Ic}}}$, η_σ and η_a are the coefficients of variation of K_{Ic}, σ and a, respectively. Substituting Eqn. (46) into Eqn. (37), it follows that

$$P_{\mathrm{f}} < \frac{\eta_{K_{\mathrm{Ic}}}^2 + \eta_\sigma^2 + \frac{1}{4}\eta_a^2}{[\ln(\mu_{K_{\mathrm{Ic}}}/\alpha\mu_\sigma\sqrt{\pi\mu_a})]^2 + \eta_{K_{\mathrm{Ic}}}^2 + \eta_\sigma^2 + \frac{1}{4}\eta_a^2} \tag{47}$$

When Eqn. (41) is used instead of Eqn. (37), we can obtain

$$P_{\mathrm{f}} < \frac{4}{9} \cdot \frac{\eta_{K_{\mathrm{Ic}}}^2 + \eta_\sigma^2 + \frac{1}{4}\eta_{a^2}^2}{[\ln(\mu_{K_{\mathrm{Ic}}}/\alpha\mu_\sigma\sqrt{\pi\mu_a})]^2} \tag{48}$$

Equations (47) and (48) are shown in Fig. 11. If K_{Ic}, σ and a follow log-normal distributions, P_{f} can be given by

$$P_{\mathrm{f}} = 1 - \Phi\left\{\frac{\ln[(\mu_{K_{\mathrm{Ic}}}/\alpha\mu_\sigma\sqrt{\pi\mu_a})((1+\eta_a^2)^{1/4}(1+\eta_\sigma^2)^{1/2}/(1+\eta_{K_{\mathrm{Ic}}}^2)^{1/2})]}{[\ln\{(1+\eta_a^2)^{1/4}(1+\eta_{K_{\mathrm{Ic}}}^2)(1+\eta_\sigma^2)\}]^{1/2}}\right\} \tag{49}$$

where $\Phi\{\ \}$ denotes the standard normal distribution function. Equation (49) is also shown in Fig. 11 to show the difference between the upper bound of P_{f} and the exact value of P_{f}.

In the preceding application, the upper bound for the probability of failure was calculated for $P_{\mathrm{f}} = \Pr(K_{\mathrm{I}} < K_{\mathrm{Ic}})$. Calculations for other fracture mechanics criteria may be conducted in a similar manner.

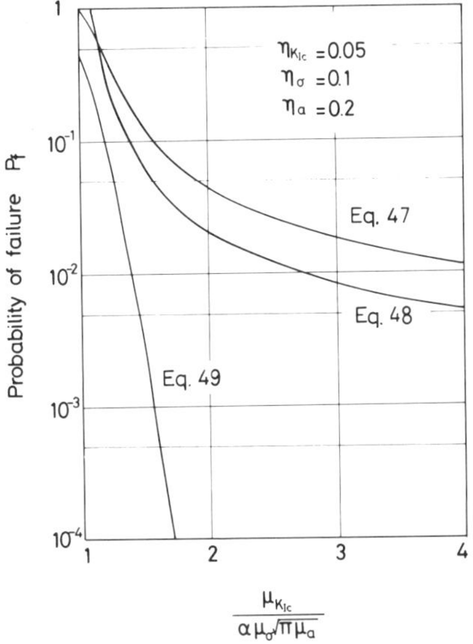

FIG. 11. Upper bound of $P_f = \Pr(K_{Ic} < \alpha\sigma\sqrt{\pi a})$.

SUMMARY

(1) Equations (5) and (12) were derived theoretically for the initial flaw size distribution and the defect detection probability, respectively. These equations are in accordance with the empirical relationships.

(2) The statistical property of the fatigue crack propagation life predicted by the composite variability model was in good agreement with experimental values.

(3) The upper bound for the probability of failure $P_f = \Pr(R < S)$ was derived using only the means and variance, and applied to a fracture mechanics problem.

ACKNOWLEDGEMENTS

One of the authors (M. I.) wishes to thank Dr S. Tanaka, President of the University of Electro-Communications, for his encouragement in this study. Thanks are also due to S. Akita and T. Takamatsu for useful discussions. This study was partly supported by a grant-in-aid for scientific research from the Japanese Ministry of Education, Science and Culture.

REFERENCES

[1] M. Ichikawa, *Reliability Eng.*, **9,** 221 (1984).
[2] M. Ichikawa, *Reliability Eng.*, **10,** 175 (1985).
[3] M. Ichikawa and S. Akita, *J. Soc. Mater. Sci. Jap.*, **35,** 1272 (1986) (in Japanese).
[4] S. Akita and M. Ichikawa, *Int. J. Fracture,* **32,** R3 (1986).
[5] M. Ichikawa, *Reliability Eng.*, **5,** 173 (1983).
[6] M. Ichikawa, *Reliability Eng.*, **8,** 57 (1984).
[7] A. L. Chandler and T. Chapman, ASME Paper, 83-GT-237 (1986).
[8] K. Arin, ASME Paper, 82-GT-324 (1986).
[9] C. H. Cook and C. F. Spaeth, ASME Paper, 82-GT-311 (1986).
[10] A. Brücker and D. Munz, *Reliability Eng.*, **5,** 139 (1983).
[11] R. Nakamura, *J. Jap. Soc. Mech. Engrs.*, **85,** 70 (1982) (in Japanese).
[12] T. Nakatsuji, M. Kuramochi, T. Fujimori, K. Sato and M. Toyoda, *J. Jap. Soc. Nondestructive Inspection,* **30,** 257 (1981) (in Japanese).
[13] M. Ichikawa, *Statistical Research on Fatigue and Fracture,* T. Tanaka, S. Nishijima and M. Ichikawa, Eds., Current Japanese Materials Research, Vol. 2, Elsevier Applied Science, London, 71 (1986).
[14] Y. Kimura and T. Kunio, *J. Soc. Mater. Sci. Jap.*, **32,** 1144 (1983) (in Japanese).
[15] H. Ishikawa and A. Tsurui, *Trans. Jap. Soc. Mech. Engrs.*, **A-50,** 1309 (1984) (in Japanese).
[16] M. D. Thien and M. Massoud, *J. Eng. Ind., Trans. ASME,* **B39,** 853 (1974).
[17] J. Hamanaka, Y. Hashimoto, M. Ito and N. Yatanabe, *Trans. Jap. Soc. Mech. Engrs.*, **A-52,** 2187 (1986) (in Japanese).
[18] M. Ichikawa and T. Takamatsu, *Proceedings of 1st Japan Conference on Structural Safety and Reliability,* Vol. 1, The Society of Materials Science, Japan, Kyoto, 341 (1987).

Practical Implementation of Fatigue-Proof Design in Structural Components with the Aid of Reliability-Based Scatter Factors

MASANOBU SHINOZUKA

Department of Civil Engineering and Operations Research, Princeton University, Princeton, New Jersey, USA

HIROSHI ISHIKAWA

Department of Information Science, Kagawa University, Takamatsu City, Kagawa, Japan

and

HIDETOSHI ISHIKAWA

Department of Information Systems, Kagawa Technical College, Marugame City, Kagawa, Japan

ABSTRACT

A definition of scatter factor is introduced that is rational and, at the same time, can be directly related to the reality of aircraft design and certification, as well as to full-scale and coupon fatigue tests of structural elements or components. Specifically, the scatter factor is defined as the ratio of the maximum likelihood estimate (MLE) of the scale parameter of the two-parameter Weibull distribution that is assumed to describe the life distribution of structural elements or components to the 'time to first failure' among a fleet of nominally identical elements or components that are also subjected to nominally identical operation conditions. Freudenthal has used the same definition of the scatter factor, although under much-simplified conditions: he assumes that the shape parameter of the Weibull distribution is known. This assumption is mathematically very convenient since it permits the distribution of the scatter factor to be derived in closed form and independent of the unknown scale parameter. Unfortunately, however, such an assumption is inconsistent with reality where the Weibull shape parameter easily ranges from 2·0 to 10·0, reflecting the fact that structural elements or components suffer from a variety of sources of randomness in fatigue strength, not only from the probabilistic variation of the material properties but also from the statistical variation in workmanship associated with, for example, drilling rivet holes in the process of airframe fabrication. The mathematical difficulty, however, multiplies when the Weibull shape and scale parameters are both assumed to be unknown. Procedures involving Monte Carlo techniques are established to evaluate the scatter factor under these conditions, using the maximum likelihood estimates of the parameters. The fleet reliability can then be estimated on the basis of the scatter factor thus evaluated. The effect of the size to be used in the fatigue test, of the fleet size and of the reliability level on the accuracy of such an estimation is also discussed.

INTRODUCTION

In a recent study [1] dealing with reliability-based criteria for design, inspection and fleet management of USAF aircraft, an effort has been made to incorporate into the reliability evaluation scheme the direct or indirect effect of material selection, geometrical configuration (in terms of either a fail-safe or slow crack growth model), mission spectra, inspection procedures, proof load test, design practice and analysis method. This study has not only provided an adequate analytical framework for the current effort toward implementing reliability-based criteria, but has also identified a number of difficulties which should be alleviated, if not totally removed, by means of a continued study, before reliability-based criteria can be accepted and implemented with sufficient engineering confidence. Among the items for which the study recommended a continued investigation was the scatter factor, which has been part of the conventional design procedure and, at the same time, is closely related to the reliability of aircraft structures as recently demonstrated by Freudenthal [2].

The general framework for developing reliability-based criteria for aircraft must emerge as a compromise among the applicability of rigorous analytical procedures, the availability of pertinent data, and the requirements for ready implementation of such criteria. These requirements, therefore, should provide a format that makes it reasonably simple to translate design procedures and design-decision processes currently in use in the U.S. Air Force and in the airframe industry into formally not-too-dissimilar procedures and processes, which should reflect the new probabilistic concept of engineering reality. The scatter factor, being probably the only quantity based on the full-scale test, plays a crucial role in certification and reliability demonstration procedures.

In the study performed here, a definition is introduced of the scatter factor that is rational and, at the same time, can be directly related to the reality of aircraft design and certification, as well as to full-scale and coupon fatigue tests of structural elements or components. Specifically, the scatter factor will be defined as the ratio of the maximum likelihood estimate (MLE) of the scale parameter of the two-parameter Weibull distribution that is assumed to describe the life distribution of structural elements or components to the 'time to first failure' among a fleet of nominally identical elements or components that are also subjected to nominally identical operating conditions. Freudenthal, in Ref. 2, has used the same definition of the scatter factor, although under much simplified conditions: he assumes that the shape parameter of the Weibull distribution is known. This assumption is mathematically very expedient since it permits the distribution of the scatter factor to be derived in closed form and independent of the unknown scale parameter as demonstrated in Ref. 2. Unfortunately, however, such an assumption is inconsistent with reality where the Weibull shape parameter easily ranges from 2·0 to 10·0, reflecting the fact that structural elements or components suffer from a variety of sources of randomness in fatigue strength, not only from the probabilistic variation of the material properties but also from the statistical variation in workmanship associated with, for example, drilling rivet holes in the process of airframe fabrication.

The mathematical difficulty multiplies when the Weibull shape and scale parameters are both assumed to be unknown [3]. The proposed Monte Carlo simulation approach can overcome the difficulty and produce results amenable to practical applications.

STATISTICAL SCATTER FACTOR WITH A KNOWN WEIBULL SHAPE PARAMETER

Recently, Freudenthal published a paper [2] proposing a statistical interpretation of the scatter factor for use in the reliability assessment of aircraft structures. The essence of his paper is briefly described here and a new interpretation is given.

Consider a two-parameter Weibull distribution for the random life (T) of a structure (or a structural component):

$$F_T(t) = 1 - \exp\left[-\left(\frac{t}{\beta}\right)^{\alpha}\right] \tag{1}$$

where it is assumed that the shape parameter (α) is known while the scale parameter (β) is unknown. The maximum likelihood estimate $\hat{B}$ (MLE) for β is given by

$$\hat{B} = \left[\frac{1}{n}\sum_{i=1}^{n} T_{0i}^{\alpha}\right]^{1/\alpha} \tag{2}$$

with $T_{01}, T_{02}, \ldots, T_{0n}$ indicating a random sample of size n taken from the distribution given in Eqn. (1).

Let T_1 represent the time to first failure (minimum life) in a fleet of m aircraft (fleet size $= m$). Then, the distribution function of T_1 is given by

$$F_{T_1}(t_1) = 1 - \exp\left[-\left(\frac{t_1}{\beta_1}\right)^{\alpha}\right] \tag{3}$$

where

$$\beta_1 = \beta / m^{1/\alpha} \tag{4}$$

Freudenthal defines the scatter factor S as

$$S = \hat{B} / T_1 \tag{5}$$

and shows that the distribution of S is given by

$$F_S(s) = \left[\frac{s^{\alpha}}{(m/n) + s^{\alpha}}\right]^n \tag{6}$$

In deriving Eqn. (6), the fact has been used that $2n(\hat{B}/\beta)^{\alpha}$ is distributed as χ^2 with $2n$ degrees of freedom, and hence the density function of $\hat{B}$ is given by

$$f_{\hat{B}}(\hat{\beta}) = \frac{n^n}{\Gamma(n)} \cdot \frac{\alpha}{\beta} \cdot \left(\frac{\hat{\beta}}{\beta}\right)^{\alpha n - 1} \exp[-n(\hat{\beta}/\beta)^{\alpha}] \tag{7}$$

It then follows that the density function $g_Z(z)$ of $Z = \hat{B}/\beta$ is such that

$$g_Z(z)\,\mathrm{d}Z = f_{\hat{B}}(\hat{\beta})\,\mathrm{d}\hat{\beta} = \frac{n^n}{\Gamma(n)} \cdot \alpha \cdot z^{\alpha n - 1} \exp[-nz^{\alpha})\,\mathrm{d}z \tag{8}$$

Consider, at this point, the following transformations:

$$V = (\hat{B}/\beta)^{\alpha} = \frac{1}{n}\sum_{i=1}^{n} (T_{0i}/\beta)^{\alpha} \tag{9}$$

and

$$W = (T_1/\beta)^{\alpha} \tag{10}$$

Note that $(T_{0i}/\beta)^\alpha$ is exponentially distributed with a unit mean value (indeed, Eqn. (7) follows from this fact) and that $(T_1/\beta)^\alpha$ is also exponentially distributed, but with a mean value equal to $1/m$. Therefore, it is easy to generate sample pairs of V and W by Monte Carlo techniques and, at the same time, compute the sample values of the ratio V/W from these. This ratio can be written from Eqns. (9) and (10) as

$$V/W = (\hat{B}/T_1)^\alpha \tag{11}$$

and hence

$$S = \hat{B}/T_1 = (V/W)^{1/\alpha} \tag{12}$$

Since α has been assumed to be known, sample values of the scatter factor (S) can be generated by means of Eqn. (12). The use of such a Monte Carlo simulation with the size of a (simulated) sample as large as 999 has yielded a satisfactory result in producing an empirical distribution of the scatter factor for the case of $m = 3$, $n = 1$ and $\alpha = 4{\cdot}0$ as shown in Fig. 1.

Although the agreement just observed between the simulation and the theory is for a particular set of parameter values, i.e. $m = 3$, $n = 1$

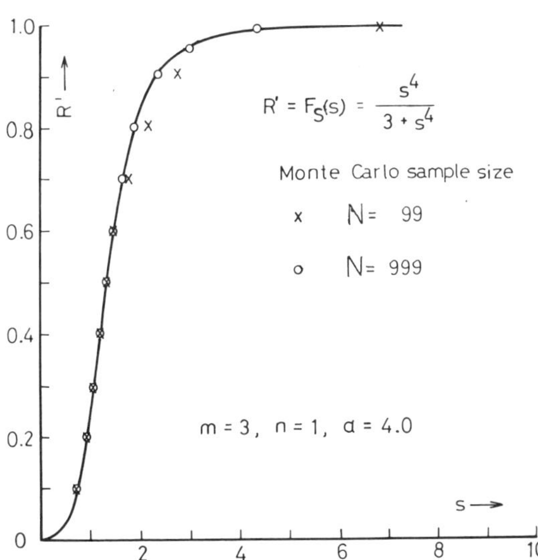

FIG. 1. Distribution function of the scatter factor.

and $\alpha = 4{\cdot}0$, it is expected that a sample size in the order of magnitude of 1000 will be sufficiently large for the type of simulations to be performed later, in which no theoretical distribution is available for comparison purposes.

Define now t_1^* as the service life specified for the fleet, or equivalently as the specified minimum life in the fleet. Define also the fleet reliability R as

$$R = P\{T_1 \geq t_1^*\} \tag{13}$$

which indicates the probability that the time to first failure in the fleet will be larger than or equal to the specified value. In Ref. 2, however, Freudenthal has implied that the probability

$$R_0' = P\{S \leq \hat{\beta}_0/t_1^*\} \tag{14}$$

should be used for the fleet reliability, where $\hat{\beta}_0$ now indicates a realization of random variable $\hat{B}$, that is a particular estimate for MLE of β for a particular sample of fatigue lives $t_{01}, t_{02}, \ldots, t_{0n}$. With the aid of Eqns. (3) and (4), the fleet reliability can be written as

$$R = \exp[-m(t_1^*/\beta)^\alpha] \tag{15}$$

while R_0' in Eqn. (14), with the aid of Eqn. (6), becomes

$$R_0' = \left[\frac{(\hat{\beta}_0/t_1^*)\alpha}{(m/n) + (\hat{\beta}_0/t_1^*)^\alpha}\right]^n \tag{16}$$

Obviously R and R_0' are not identical. The difference can be more clearly demonstrated by rewriting Eqns. (13) and (14), respectively, as

$$R = \int_{D_1 \cup D_2}\!\!\int f_{T_1}(t_1) f_{\hat{B}}(\hat{\beta})\, \mathrm{d}t_1\, \mathrm{d}\hat{\beta} \tag{17}$$

$$R_0' = \int_{D_1 \cup D_3}\!\!\int f_{T_1}(t_1) f_{\hat{B}}(\hat{\beta})\, \mathrm{d}t_1\, \mathrm{d}\hat{\beta} \tag{18}$$

where $\hat{\beta}$ has been introduced to represent the realization of $\hat{B}$ in general. The domains of integration $D_1 \cup D_2$ (D_1 union D_2) and $D_1 \cup D_3$ are shown in Fig. 2. The domain $D_1 \cup D_3$ in Eqn. (18) can be obtained by adding D_3 to and subtracting D_2 from the correct domain of integration $D_1 \cup D_2$. Note that D_3 represents a conservative addition and D_2 an unconservative subtraction. In Eqns. (17) and (18), $f_{T_1}(t_1)$ represents the density function of T_1.

Since $\hat{\beta}_0$ is a particular realization for MLE $\hat{B}$ of β, it will assume

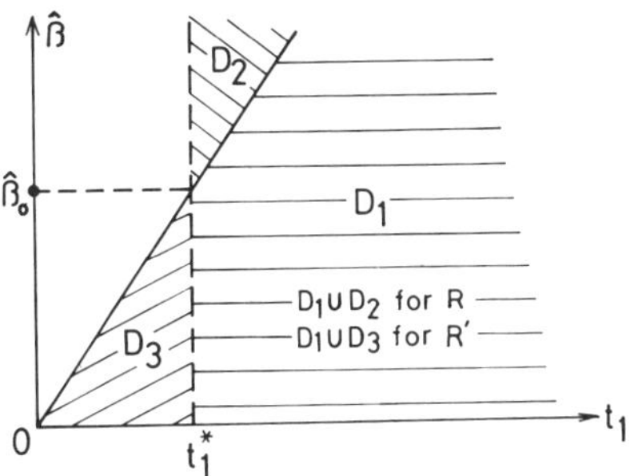

FIG. 2. Domains of integration.

different values of fatigue life from sample to sample. Therefore, the necessity arises from time to time to interpret $\hat{\beta}_0$ as a random variable $\hat{B}_0$ with a density function $f_{\hat{B}_0}(\hat{\beta})$ identical with that of $\hat{B}$ as shown in Eqn. (7). Therefore, if $\hat{B}$ is used in place of β_0 in Eqn. (16), we can construct a random variable R', the particular realization of which is R_0';

$$R' = \left[\frac{(\hat{B}/t_1^*)^\alpha}{(m/n) + (\hat{B}/t_1^*)^\alpha}\right]^n \tag{19}$$

To demonstrate that R' may be used as an approximation for R, consider the expected value $E[(R')^k]$ of $(R')^k$;

$$E[(R')^k] = \int_0^\infty (R')^k f_{\hat{B}}(\hat{\beta})\, \mathrm{d}\hat{\beta} \tag{20}$$

where R' and $f_{\hat{B}}(\hat{\beta})$ in the integrant of the second member are to be replaced by the right-hand sides of Eqns. (19) (with $\hat{B}$ being replaced by $\hat{\beta}$) and (7), respectively. The integration then depends on t_1^*, β, m and n. The first three of these quantities are related, however, in the following fashion through Eqn. (15):

$$(t_1^*/\beta)^\alpha = -(\ln R)/m \tag{21}$$

With the aid of Eqn. (21), t_1^* and m in Eqn. (19) can be eliminated and, by defining $Z = \hat{B}/\beta$, Eqn. (19) becomes

$$R' = \left[\frac{(\hat{B}/\beta)^\alpha}{(\hat{B}/\beta)^\alpha - (\ln R)/n}\right]^n = \left[\frac{Z^\alpha}{Z^\alpha - (\ln R)/n}\right]^n \tag{22}$$

Substituting Eqns. (8) and (22) (with Z being replaced by z) into

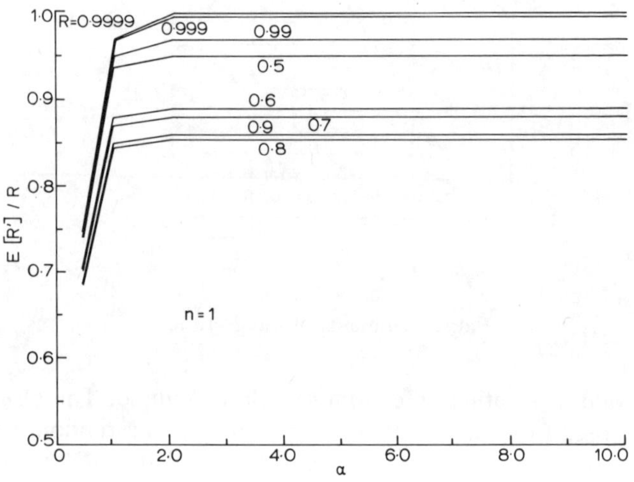

FIG. 3. $E[R']/R$ as a function of α ($n = 1$).

Eqn. (20), one can obtain

$$E[(R')^k] = \int_0^\infty \left[\frac{nz^\alpha}{nz^\alpha - \ln R}\right]^{kn} \frac{\alpha n^n}{\Gamma(n)} z^{\alpha n - 1} \exp[-nz^\alpha]\, \mathrm{d}z \qquad (23)$$

The mean $E[R']$ and the standard deviation $\sigma_{R'}$ of R' can easily be obtained from Eqn. (23).

The integrals in Eqn. (23) with $k = 1$ and 2, which are now independent of m, were numerically evaluated for $\alpha = 0{\cdot}5$, $1{\cdot}0$, $2{\cdot}0, \ldots, 10{\cdot}0$, $R = 0{\cdot}5, 0{\cdot}6, \ldots, 0{\cdot}9, 0{\cdot}99, 0{\cdot}999, 0{\cdot}9999$, and $n = 1, 2, \ldots, 10$. The ratios $E[R']/R$ for these values of R are plotted here as functions of α in Figs. 3 and 4 for $n = 1$ and 10, respectively. Also, the coefficients of variation $V_{R'} = \sigma_{R'}/E[R']$ of R' are plotted as functions of α for the same values of R and n in Figs. 5 and 6. (Similar plots for other values of n can be found in Ref. 4.) One observes from these results the following general trend:

(1) The ratio $E[R']/R$ increases with increasing α. However, except for a sharp increase observed between $\alpha = 0{\cdot}5$ and $1{\cdot}0$, the ratio is almost constant for those values of $\alpha \geq 2{\cdot}0$, which include a practical range for α.
(2) With R fixed, the ratio is closer to unity for a larger value of n.
(3) The ratio is not necessarily larger for a large value of R,

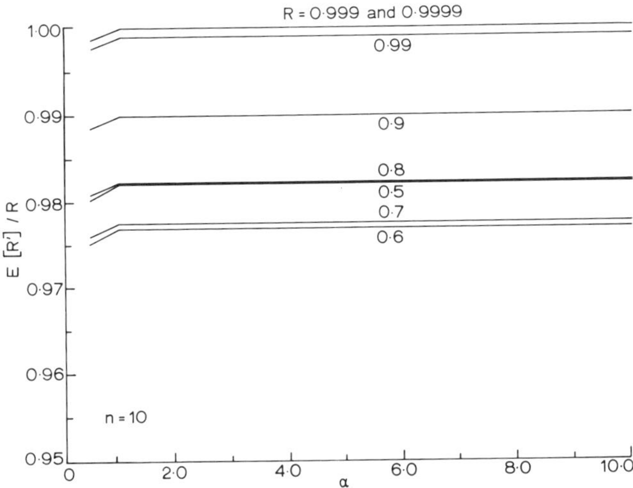

FIG. 4. $E[R']/R$ as a function of α ($n = 10$).

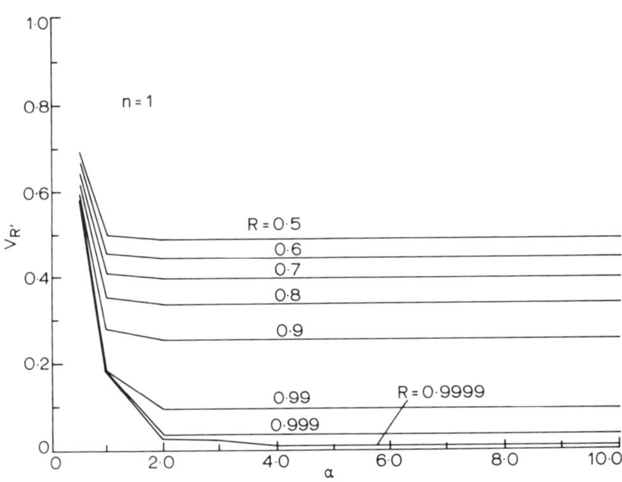

FIG. 5. $V_{R'}$ as a function of α ($n = 1$).

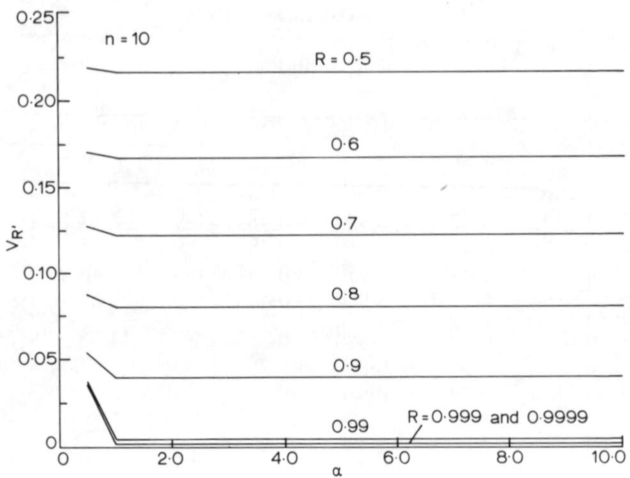

FIG. 6. $V_{R'}$ as a function of α ($n = 10$).

although numerical results show that $E[R']$ monotonically increases as a function of R with all other parameters kept constant.

(4) As α increases, the coefficient of variation $V_{R'}$ of R' decreases. Except for a sharp decrease observed between $\alpha = 0{\cdot}5$ and $2{\cdot}0$, however, the coefficient is almost constant for $\alpha \geq 2{\cdot}0$.
(5) With R fixed, the coefficients are smaller when n is larger.
(6) The coefficients are smaller for a larger value of R.

For example, the ratio $E[R']/R$ is nearly equal to 0·950 for $R = 0{\cdot}5$ and $n = 1$ (Fig. 3). This may give an impression that the approximation is acceptable. However, the fact that the corresponding coefficient of variation $V_{R'}$ is as large as 0·50 (Fig. 5) proves that this impression is unsubstantiated. As n increases to 10 (while $R = 0{\cdot}5$ is kept constant), the ratio $E[R']/R$ increases to 0·982 (Fig. 4) and, at the same time, the coefficient $V_{R'}$ decreases to 0·22 (Fig. 6), thus making the approximation more reasonable. For $R = 0{\cdot}9$, however, the approximation appears to be reasonable since the ratios are then 0·88 and 0·99, and the coefficients are 0·27 and 0·05, respectively, for $n = 1$ and 10. Table 1 summarizes such observations for selected values of R, α and n. The last column in Table 1 lists the value of Freudenthal's scatter factor corresponding to $E[R']$ for different values of m.

TABLE 1
Values of $E[R']$, $E[R']/R$, $V_{R'}$ and scatter factors for selected values of R, α and n

n	r	α	$E[R']$	$E[R']/R$	$V_{R'}$	S		
						$m=1$	$m=5$	$m=25$
1	0·50	2	0·475	0·950	0·488	0·95	2·13	4·76
		3	0·475	0·950	0·488	0·96	1·65	2·83
		4	0·475	0·950	0·488	0·97	1·46	2·18
		5	0·475	0·950	0·488	0·98	1·35	1·87
	0·90	2	0·792	0·880	0·253	1·95	4·36	9·76
		3	0·792	0·880	0·253	1·56	2·67	4·57
		4	0·792	0·880	0·253	1·40	2·09	3·12
		5	0·792	0·880	0·253	1·31	1·80	2·49
	0·99	2	0·959	0·969	0·093	4·84	10·80	24·20
		3	0·959	0·969	0·093	2·86	4·89	8·37
		4	0·959	0·969	0·093	2·20	3·29	4·92
		5	0·959	0·969	0·093	1·88	2·59	3·58
5	0·50	2	0·485	0·970	0·296	1·13	2·53	5·67
		3	0·485	0·970	0·296	1·09	1·86	3·18
		4	0·485	0·970	0·296	1·06	1·59	2·38
		5	0·485	0·970	0·296	1·05	1·45	2·00
	0·90	2	0·881	0·979	0·066	2·79	6·23	13·90
		3	0·881	0·979	0·066	1·98	3·39	5·79
		4	0·881	0·979	0·066	1·67	2.50	3·73
		5	0·881	0·979	0·066	1·51	2·08	2·87
	0·99	2	0·988	0·998	0.0072	8·93	20.00	44·70
		3	0·988	0·998	0.0072	4·31	7·36	12·60
		4	0·988	0·998	0.0072	2·99	4·47	6·68
		5	0·988	0·998	0.0072	2·40	3·31	4·57
10	0·50	2	0·491	0·982	0·216	1.16	2·60	5·82
		3	0·491	0·982	0·216	1.11	1·89	3·24
		4	0·491	0·982	0·216	1.08	1·61	2·41
		5	0·491	0·982	0·216	1.06	1·47	2·02
	0·90	2	0·891	0·990	0·040	2·93	6·56	14·70
		3	0·891	0·990	0·040	2·05	3·50	5·99
		4	0·891	0·990	0·040	1·71	2·56	3·83
		5	0·891	0·990	0·040	1·54	2·12	2·93
	0·99	2	0·989	0·999	0·004	9·47	21·20	47·30
		3	0·989	0·999	0·004	4·47	7·65	13·10
		4	0·989	0·999	0·004	3·08	4·60	6·88
		5	0·989	0·999	0·004	2·46	3·39	4·68

It is important to indicate the justification with which the Weibull distribution has been used in the analysis:

(1) The Weibull distribution (Eqn. (1)), is certainly one of the distributions we can use to fit to the central part of the empirical data. We can estimate the scale parameter $\hat{\beta}$, which is a measure of the central location of the distribution under the assumptions of Eqn. (1).
(2) The random variable T_1 represents an extreme value and can be assumed to follow the Weibull distribution (Eqn. (3)) as long as m is reasonably large (say $m \geq 5$).

STATISTICAL SCATTER FACTOR WITH UNKNOWN SHAPE AND SCALE PARAMETERS

It is well known that, when the shape and scale parameters of the two-parameter Weibull distribution given in Eqn. (1) are both unknown, the MLE $\hat{A}$ of α and MLE $\check{B}$ of β are obtained from the following simultaneous equations:

$$n = \sum_{i=1}^{n} (T_{0_i}/\check{B})^{\hat{A}} \tag{24}$$

$$n = \sum_{i=1}^{n} [(T_{0i}/\check{B})^{\hat{A}} - 1] \ln(T_{0i}/\check{B})^{\hat{A}} \tag{25}$$

where, as before, $T_{01}, T_{02}, \ldots, T_{0n}$ indicate a random sample of size n taken from the Weibull distribution. In this case, the statistical scatter factor Q is introduced in the following form as a natural extension of S given in Eqn. (5).

$$Q = \check{B}/T_1 \tag{26}$$

Defining $Y(i)$, U and V_0 as

$$Y(i) = (T_{0i}/\beta)^{\alpha} \tag{27}$$

$$U = \hat{A}/\alpha \tag{28}$$

and

$$V_0 = (\check{B}/\beta)^{\alpha} \tag{29}$$

one can rewrite Eqns. (24) and (25), respectively, in the following forms, involving their respective realizations $Y(i)$, U and V_0:

$$v_0 = \left[\sum_{i=1}^{n} y^u(i)/n \right]^{1/u} \tag{30}$$

$$\phi(u) = \frac{\sum_{i=1}^{n} y^u(i) \ln y(i)}{\sum_{i=1}^{n} y^u(i)} - \frac{1}{u} - \frac{\sum_{i=1}^{n} \ln y(i)}{n} = 0 \tag{31}$$

Since $Y(i)$ is exponentially distributed with a mean value equal to unity, the joint distribution of U and V_0 can be obtained through Eqns. (30) and (31) by the Monte Carlo simulation technique. All that one is required to do is to generate $y(i)$ $(i = 1, 2, \ldots, n)$, use them in Eqns. (30) and (31), solve these two equations for u and v_0, and repeat the process N times. This will produce a simulated sample of pairs U and V_0 of size N. Independently of the simulation of U and V_0, a (simulated) sample of W (Eqn. (10)) of size N is again generated.

By means of the Monte Carlo simulation technique just described, Whittaker and Besuner [5] constructed empirical distribution functions of U^* and V_0^* defined as

$$U^* = 1/U \tag{32}$$

$$V_0^* = V_0^U = (\check{B}/\beta)^{\hat{A}} \tag{33}$$

These empirical distributions are simulated, based on samples of size 1999, and are shown in Figs. 7 and 8, respectively. The comparison of current simulation results with those in Ref. 5 indicates that the accuracy of the current simulation is generally comparable with that of Ref. 5.

Since

$$Q^\alpha = \left(\frac{\check{B}}{T_1}\right)^\alpha = \frac{(\check{B}/\beta)^\alpha}{(T_1/\beta)^\alpha} = \frac{V_0}{W} \tag{34}$$

it follows that

$$Q^* = Q^{\hat{A}} = (V_0/W)^U \tag{35}$$

and therefore Q^* is a parameter-free statistic. The last equation indicates that the distribution of Q^* can be constructed with the aid of u, v_0 and w already generated. For ease of application, however, the distribution of the logarithm Z of Q^*,

$$Z = \log_{10} Q^* = U \log_{10}(V_0/W) \tag{36}$$

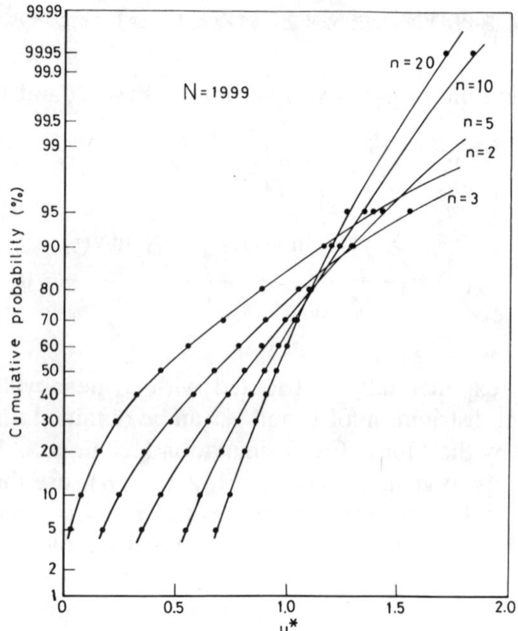

FIG. 7. Empirical distribution of u^*.

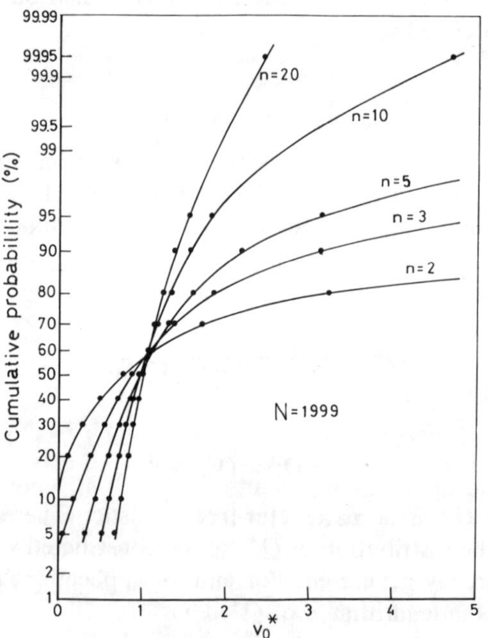

FIG. 8. Empirical distribution of V_0^*.

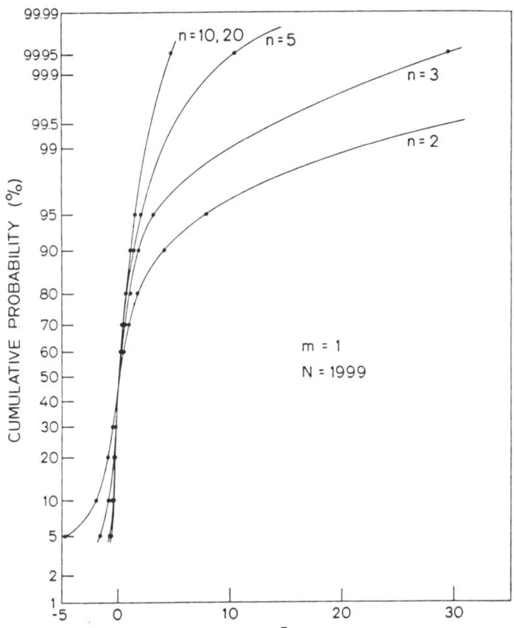

FIG. 9. Empirical distribution of Z ($m = 1$).

is constructed for $m = 1$, 3, 5, 10, 25, 100, 200 and 1000, and plotted only for $m = 1$ and 100 in Figs. 9 and 10, respectively, where solid circles indicate the values of the empirical distribution function of Z based on samples of size 1999. Solid curves are drawn through these circles by interpolation. Similarly, Fig. 11 plots the empirical distribution of Z for various values of m when $n = 3$.

Having thus established the empirical distribution of Z, define the probability R_0'' as

$$R_0'' = P\{Q^{\hat{A}} \leq (\check{\beta}_0/t_1^*)^{\hat{\alpha}_0}\} = P\{Q^* \leq (\check{\beta}_0/t_1^*)^{\hat{\alpha}_0}\} \tag{37}$$

where t_1^* is the specified minimum life, and $\hat{\alpha}_0$ and $\check{\beta}_0$ represent particular realizations of the MLE $\hat{A}$ and $\check{B}$ based on a sample of size n taken from Eqn. (1) of $t_{01}, t_{02}, \ldots, t_{0n}$. This probability R_0'' is similar to Freudenthal's fleet reliability R_0' defined by Eqn. (14), and can be expressed, with the aid of Eqn. (36), as

$$R_0'' = P\{Z \leq \log_{10}(\check{\beta}_0/t_1^*)^{\hat{\alpha}_0}\} \tag{38}$$

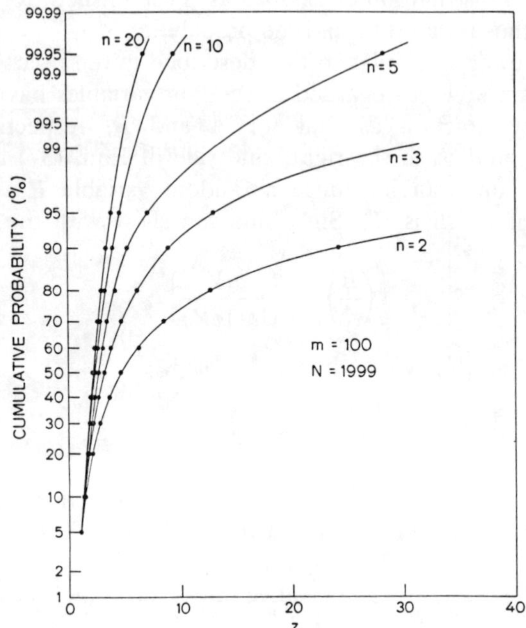

FIG. 10. Empirical distribution of Z ($m = 100$).

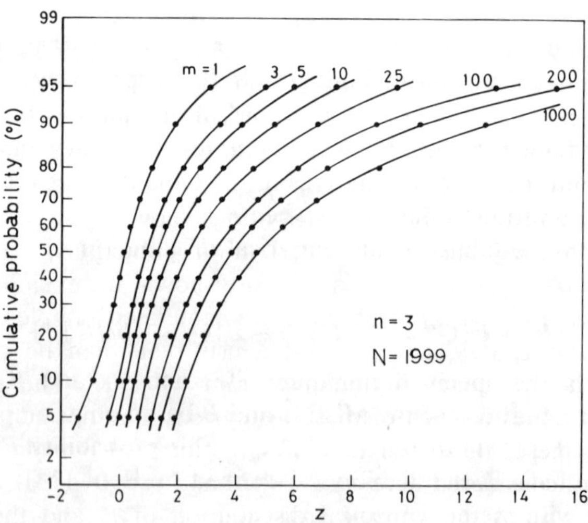

FIG. 11. Empirical distribution of Z ($n = 3$).

The empirical distribution of Z, such as is established, for example, in Fig. 9, can thus be used to find the probability R''_0.

From a viewpoint similar to that described in the preceding section, $\hat{\alpha}_0$ and $\check{\beta}_0$ can now be regarded as random variables having the same joint density function as that for $\hat{A}$ and $\check{B}$, respectively. Then, replacing $\hat{\alpha}_0$ and $\check{\beta}_0$ in the right-hand side of Eqn. (37) by $\hat{A}$ and $\check{B}$, respectively, one can introduce a random variable R'', a particular realization of which is R''_0. Since one can show with the aid of Eqn. (21) that

$$\left(\frac{\check{B}}{t_1^*}\right)^{\hat{A}} = \left\{\frac{mV_0}{\ln(1/R)}\right\}^U \tag{39}$$

the random variable R'' can be written as

$$R'' = P\left[Z \leq U \log_{10}\left\{\frac{mV_0}{\ln(1/R)}\right\}\right] = F_Z\left[U \log_{10}\left\{\frac{mV_0}{\ln(1/R)}\right\}\right] \tag{40}$$

where $F_Z[\cdot]$ is the (cumulative) distribution function of Z. In the computation that follows, empirical distributions such as those obtained in Figs. 9 and 10 were used in place of $F_Z[\cdot]$. The expected values of $(R'')^k$, where k is a positive integer, can then be evaluated as

$$E[(R'')^k] = \int_0^\infty \int_0^\infty F_Z^k\left[u \log_{10}\left\{\frac{mv_0}{\ln(1/R)}\right\}\right] \cdot f_{UV_0}(u, v_0)\, du\, dv_0 \tag{41}$$

Obviously, the expected value of R'' is obtained as $E[R'']$, and the variance of R'' as $\text{Var}[R''] = E[(R'')^2] - E^2[R'']$. In Eqn. (41), $f_{UV_0}(\cdot, \cdot)$ is the joint density function of U and V_0, which can be empirically evaluated by making use of Eqns. (30) and (31). In fact, the random variables $U^* = 1/U$ and $V_0^* = V_0^U$ considered before have a joint distribution resulting from the same underlying dependence as that for U and V_0, and the empirical distributions shown in Figs. 7 and 8 are their respective marginal distribution functions. The (two-dimensional) histograms of U and V_0 are shown, for example, in Table 2 for the case of $n = 3$ on the basis of a simulated sample of $N = 1999$ pairs of U and V_0. The empirical density can then be obtained by dividing the frequency in the histogram by 1999. The sum of the frequency values in each of these histograms may not add up to 1999 because some of the (simulated) observations fell outside the domain of u and v_0 considered here of $u = 0$–20 and $v_0 = 0$–4.

Making use of the empirical distribution of Z and the empirical density of U and V_0, one can evaluate Eqn. (41) for $k = 1$ and 2, and

TABLE 2

Two-dimensional histogram of u and v_0 ($n = 3$)

u \ v_0	*1*	*2*	*3*	*4*	*5*	*6*	*7*	*8*	*9*	*10*	*11*	*12*	*13*	*14*	*15*	*16*
1	1	6	3	1	1	1	0	0	0	0	0	0	0	0	0	0
2	30	104	89	82	59	48	19	16	5	4	2	0	2	1	0	0
3	22	64	96	102	90	61	44	17	17	14	4	0	2	2	1	0
4	11	46	68	54	57	33	33	21	10	11	6	3	1	0	1	0
5	5	25	21	29	34	18	14	12	3	6	3	0	1	0	0	0
6	5	11	20	9	22	12	11	8	4	3	1	1	1	0	1	1
7	3	13	27	14	12	4	13	5	4	1	0	0	0	0	0	0
8	3	8	11	12	5	8	3	3	1	3	1	0	1	0	0	0
9	1	4	10	10	6	3	3	2	2	0	0	1	0	0	0	0
10	2	11	3	4	8	1	1	0	1	0	0	0	0	0	0	0
11	1	1	1	3	3	3	1	1	0	2	0	0	0	0	0	0
12	0	1	2	5	3	2	2	1	0	0	0	0	0	0	0	0
13	1	2	3	3	1	2	1	0	1	0	1	0	0	0	0	0
14	0	1	2	2	3	1	1	0	0	0	0	0	0	1	0	0
15	0	0	2	2	1	0	1	0	0	0	0	0	0	0	0	0
16	0	0	1	2	1	0	0	1	0	0	0	0	0	0	0	0
17	0	1	1	2	1	0	0	0	0	0	0	0	0	0	0	0
18	0	0	1	1	1	0	0	0	0	0	0	0	0	0	0	0
19	0	0	1	0	2	1	0	0	0	0	0	1	0	0	0	0
20	0	0	1	0	0	0	0	0	0	0	0	0	0	0	0	0
21	0	0	1	1	0	0	0	1	0	0	0	0	0	0	0	0
22	1	0	0	2	0	0	0	0	0	0	0	0	0	0	0	0
23	0	0	0	0	1	0	0	0	1	0	0	0	0	0	0	0
24	0	0	0	0	1	0	0	0	0	0	0	0	0	0	0	0
25	0	0	1	1	0	0	0	0	0	0	0	0	0	0	0	0
26	0	1	0	0	0	0	0	0	0	0	0	0	0	0	0	0
27	0	1	0	1	1	0	0	0	0	0	0	0	0	0	0	0
28	0	0	1	1	0	0	0	0	0	0	0	0	0	0	1	0
29	0	1	0	0	0	0	0	0	0	0	0	0	0	0	0	0
30	0	0	0	0	0	0	0	0	0	0	0	0	0	0	0	0
31	0	0	0	1	1	0	0	0	0	0	0	0	0	0	0	0
32	0	0	0	0	1	0	0	0	0	0	0	0	0	0	0	0
33	0	3	0	0	0	0	0	0	0	0	0	0	0	0	0	0
34	0	0	0	0	0	0	0	0	0	0	0	0	0	0	0	0
35	0	0	1	0	0	0	0	0	0	0	0	0	0	0	0	0
36	0	0	0	0	0	0	0	0	0	0	0	0	0	0	0	0
37	0	0	0	0	2	0	0	0	0	0	0	0	0	0	0	0
38	0	0	0	0	0	0	0	0	0	0	0	0	0	0	0	0
39	0	0	0	0	0	0	0	0	0	0	0	0	0	0	0	0
40	0	0	0	0	0	0	0	0	0	0	0	0	0	0	1	0

Notes: For v_0, interval 1 = 0–0·25, interval 2 = 0·25–0·50, etc.
For u, interval 1 = 0–0·50, interval 2 = 0·50–1·00, etc.

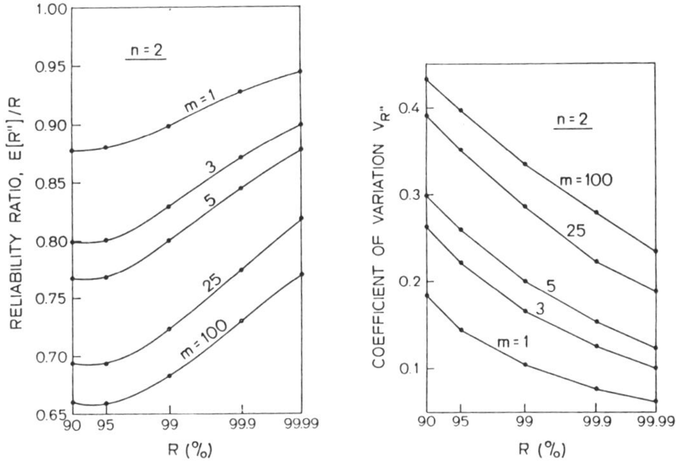

FIG. 12. $E[R'']/R$ and $V_{R''}$ as functions of R ($n = 2$).

hence find the mean value and the coefficient of variation $V_{R''}$ of R''. The results of computation for $n = 2$ and 20 are shown in Figs. 12 and 13 as a function of R, and also in Figs. 14 and 15 as a function of $E[R'']$. Similar to Table 1, Table 3 summarizes these results for selected values of R, m and n. In Table 3, however, the values of Q^* corresponding to the fleet reliability $R = E[R'']$ are listed in place of S.

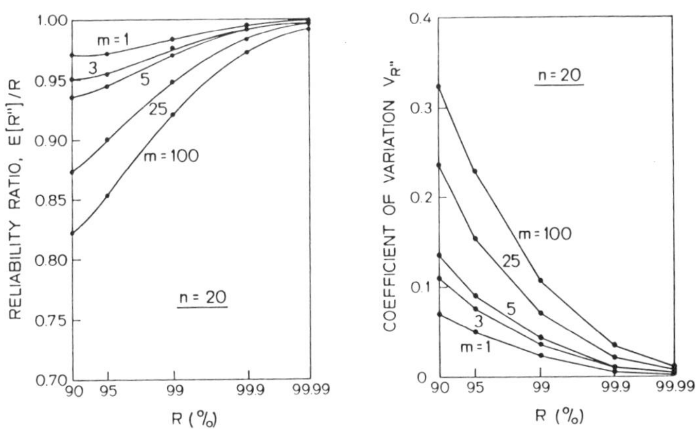

FIG. 13. $E[R'']/R$ and $V_{R''}$ as functions of R ($n = 20$).

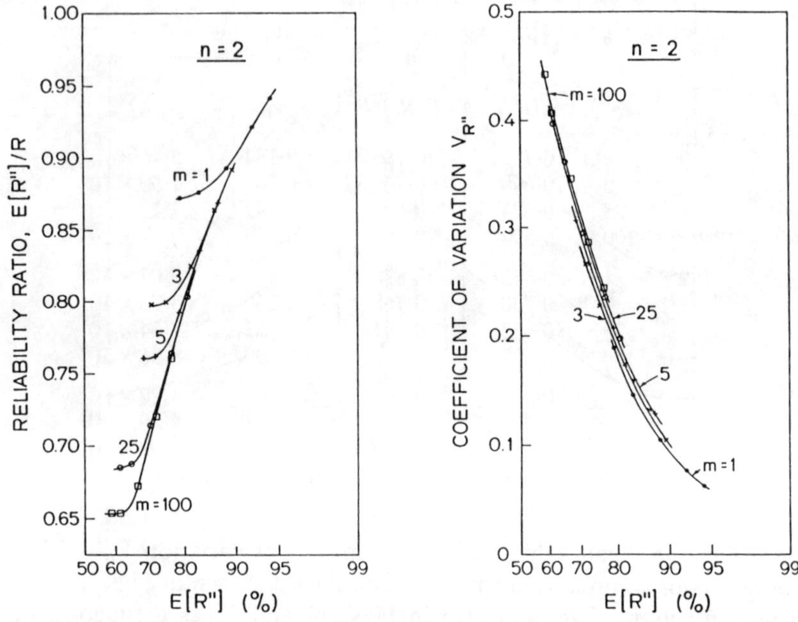

FIG. 14. $E[R'']/R$ and $V_{R''}$ as functions of $E[R'']$ ($n = 2$).

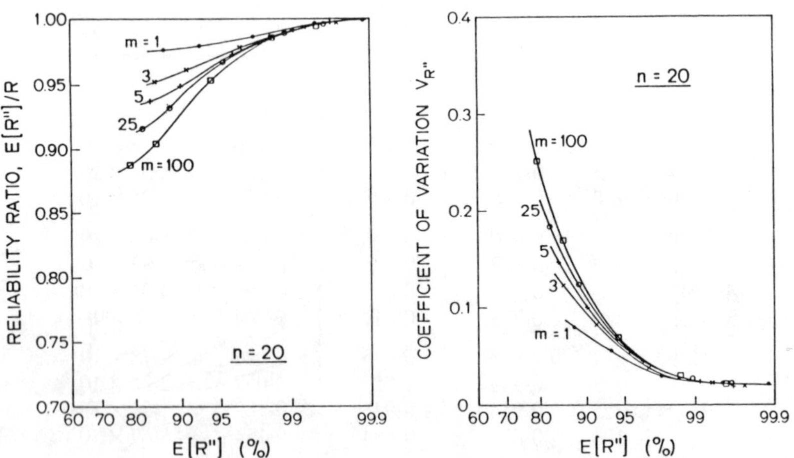

FIG. 15. $E[R'']/R$ and $V_{R''}$ as functions of $E[R'']$ ($n = 20$).

TABLE 3
Values of $E[R'']$, $E[R'']/R$, $V_{R''}$ and Q^* for selected values of R, m and n

n	R	m	$E[R'']$	$E[R'']/R$	$V_{R''}$	Q^*
2	0·90	1	0·789	0·786	0·184	$5{\cdot}46 \times 10$
		5	0·690	0·767	0·300	$1{\cdot}90 \times 10^3$
		25	0·624	0·693	0·392	$7{\cdot}72 \times 10^4$
		100	0·594	0·660	0·435	$2{\cdot}07 \times 10^6$
	0·95	1	0·835	0·879	0·144	$2{\cdot}07 \times 10^2$
		5	0·730	0·768	0·260	$5{\cdot}73 \times 10^3$
		25	0·657	0·691	0·353	$2{\cdot}54 \times 10^5$
		100	0·622	0·655	0·399	$8{\cdot}96 \times 10^6$
	0·99	1	0·888	0·897	0·103	$4{\cdot}20 \times 10^3$
		5	0·791	0·799	0·201	$8{\cdot}49 \times 10^4$
		25	0·716	0·723	0·286	$2{\cdot}98 \times 10^6$
		100	0·675	0·681	0·335	$7{\cdot}81 \times 10^7$
5	0·90	1	0·844	0·973	0·126	$1{\cdot}31 \times 10$
		5	0·777	0·863	0·217	$7{\cdot}45 \times 10$
		25	0·710	0·789	0·313	$4{\cdot}61 \times 10^2$
		100	0·669	0·743	0·370	$2{\cdot}76 \times 10^3$
	0·95	1	0·895	0·942	0·092	$2{\cdot}62 \times 10$
		5	0·835	0·879	0·165	$1{\cdot}44 \times 10^2$
		25	0·766	0·806	0·256	$8{\cdot}74 \times 10^2$
		100	0·721	0·759	0·316	$4{\cdot}86 \times 10^3$
	0·99	1	0·955	0·964	0·046	$1{\cdot}33 \times 10^2$
		5	0·912	0·922	0·095	$7{\cdot}10 \times 10^2$
		25	0·856	0·865	0·160	$4{\cdot}25 \times 10^3$
		100	0·808	0·816	0·220	$1{\cdot}96 \times 10^4$
10	0·90	1	0·867	0·963	0·096	9·52
		5	0·819	0·910	0·166	$4{\cdot}73 \times 10$
		25	0·762	0·847	0·258	$2{\cdot}34 \times 10^2$
		100	0·718	0·798	0·334	$9{\cdot}05 \times 10^2$
	0·95	1	0·920	0·968	0·067	$1{\cdot}68 \times 10$
		5	0·877	0·923	0·125	$8{\cdot}83 \times 10$
		25	0·823	0·867	0·196	$4{\cdot}35 \times 10^2$
		100	0·778	0·819	0·264	$1{\cdot}70 \times 10^3$
	0·99	1	0·973	0·983	0·026	$9{\cdot}78 \times 10$
		5	0·949	0·959	0·057	$2{\cdot}96 \times 10^2$
		25	0·914	0·923	0·102	$1{\cdot}58 \times 10^3$
		100	0·875	0·884	0·154	$6{\cdot}67 \times 10^3$

(*continued*)

TABLE 3—*contd.*

n	R	m	$E[R'']$	$E[R'']/R$	$V_{R''}$	Q^*
20	0·90	1	0·875	0·972	0·069	9·17
		5	0·842	0·935	0·134	$4·05 \times 10$
		25	0·786	0·873	0·233	$1·92 \times 10^2$
		100	0·739	0·822	0·324	$7·28 \times 10^2$
	0·95	1	0·921	0·970	0·050	$1·87 \times 10$
		5	0·897	0·945	0·090	$7·30 \times 10$
		25	0·856	0·901	0·155	$3·20 \times 10^2$
		100	0·810	0·853	0·230	$1·12 \times 10^3$
	0·99	1	0·974	0·984	0·025	$6·71 \times 10$
		5	0·961	0·971	0·042	$2·84 \times 10^2$
		25	0·940	0·949	0·070	$1·08 \times 10^3$
		100	0·912	0·921	0·107	$3·58 \times 10^3$

The following observations can be made from these results.

(1) The ratio $E[R'']/R$ increases as n and R increase, but decreases as m increases. For the ranges of n, R and m examined here, the smallest (worst) is 0·655 ($n = 2$, $R = 0·95$ and $m = 100$) while the largest (best) is 0·995 ($n = 20$, $R = 0·999$ and $m = 1$). This indicates that $E[R'']$ and R are essentially of the same order of magnitude in these ranges of n, R and m.

(2) The coefficient of variation $V_{R''}$ decreases as n and R increase, while it increases as m increases.

(3) Figures 12 and 13 are useful in practical applications. Realizations $\hat{\alpha}_0$ and $\check{\beta}_0$ of $\hat{A}$ and $\check{B}$, respectively, on a sample of size n can be used in Eqn. (38) to obtain a realization R''_0 of R''. Actually, this is done by reading the value of R''_0 from, for example, Figs. 9 or 10. Diagrams showing $V_{R''}$ as a function of $E[R'']$ in Figs. 14 and 15 then indicate whether R''_0 just obtained can be used for $E[R'']$, depending on the fleet size m and the sample size n: if $V_{R''}$ corresponding to $E[R''] = R''_0$ is small (say, less than 0.20) for m and n under consideration, $E[R'']$ may indeed be considered equal to R''_0 in approximation. Using this value of $E[R'']$ in the diagrams showing $E[R'']/R$ as a function of $E[R'']$ in Figs. 12 and 13, we can estimate the corresponding value of R. As this process for evaluating R suggests, Figs. 12 and 13 implicitly indicate for which combinations of m, n and

the specified value of $V_{R''}$ the probability R''_0 can be used as an approximation to R. If the value of $V_{R''}$ corresponding to R''_0 (replacing $E[R'']$ in Figs. 12 and 13) exceeds the specified value of $V_{R''}$ for a particular set of m and n, then one or any combination of the following should be implemented to make R''_0 a better approximation to R:

(a) The fleet size m to be considered should be reduced. This requires a re-evaluation of R''_0.

(b) The sample size n should be increased. This requires testing an additional number of structural components and a re-evaluation of R''_0 based on that.

(c) The specified minimum life t_1^* should be reduced. This requires a re-evaluation of R''_0.

(4) Figures 14 and 15 can also be used for similar purposes where $E[R'']/R$ and $V_{R''}$ are plotted as functions of R.

The foregoing work was done in order to establish methods for estimating the fleet reliability R by extending the notion of the familiar scatter factor into that of the statistical scatter factor. For this reason, it is expected that the estimation methods developed will be more attractive and more readily acceptable by the profession than other possible alternatives. For example, it is possible to use the MLEs $\hat{A}$ and $\check{B}$ directly in Eqn. (15) to obtain $\check{R}$ as an estimator for R.

$$\check{R} = \exp\left[-m\left(\frac{t^*}{\check{B}}\right)^{\hat{A}}\right] = \exp\left[-m\left(-\frac{\ln R}{m}\right)^{U} \cdot V_0^{-U}\right] \tag{42}$$

The expected value and coefficient of variation of $\check{R}$ can also be evaluated with the aid of the empirical joint distribution of U and V_0. Figures 16–18 give some of simulated results for statistical behavior of $\check{R}$. In these figures, the corresponding results for R'' are also illustrated for comparison purposes. From these figures, it was found that such an estimator $\check{R}$ exhibited more favourable statistical features than R'' in the range of lower fleet reliability level. However, in the range of considerably higher fleet reliability, $\check{R}$, in addition to its lack of a direct bearing on conventional design and certification procedures, exhibited less favorable statistical features than R''; for example, the coefficients of variation were generally larger than those associated with R'' in similar ranges of fleet size, test sample size and fleet reliability.

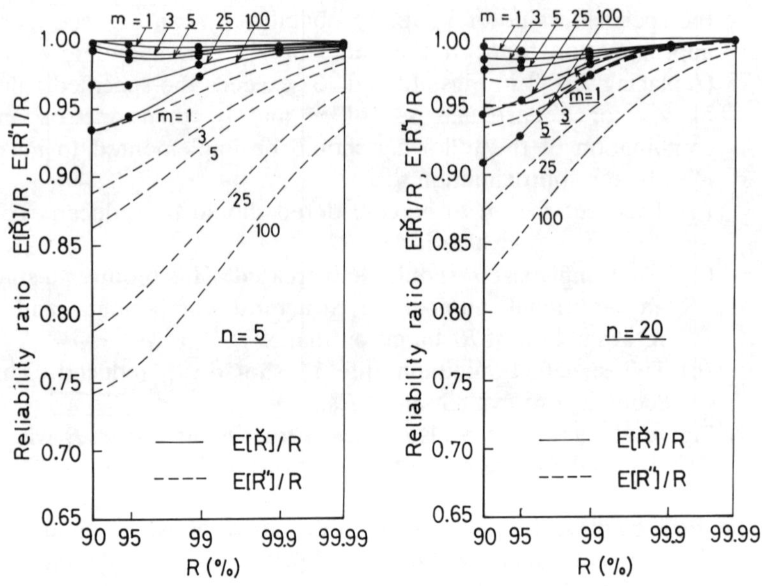

FIG. 16. Expected values of $\check{R}$ and R'' as a function of fleet reliability R ($n = 5$ and 20).

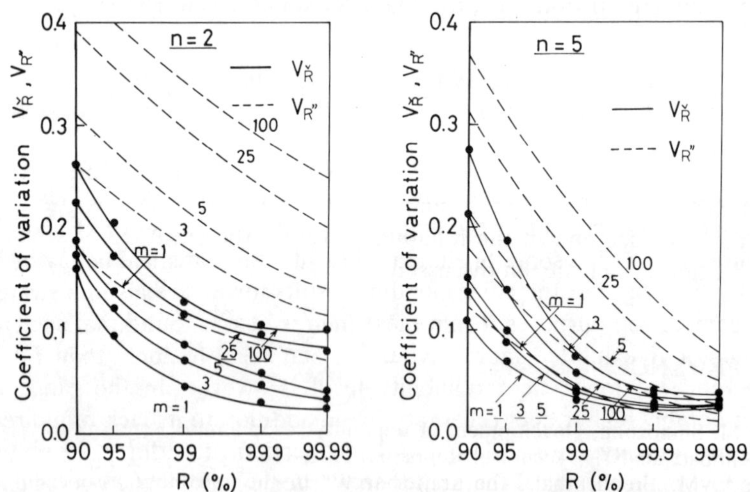

FIG. 17. Coefficients of variation, $V_{\check{R}}$ and $V_{R''}$, as a function of fleet reliability R ($n = 2$ and 5).

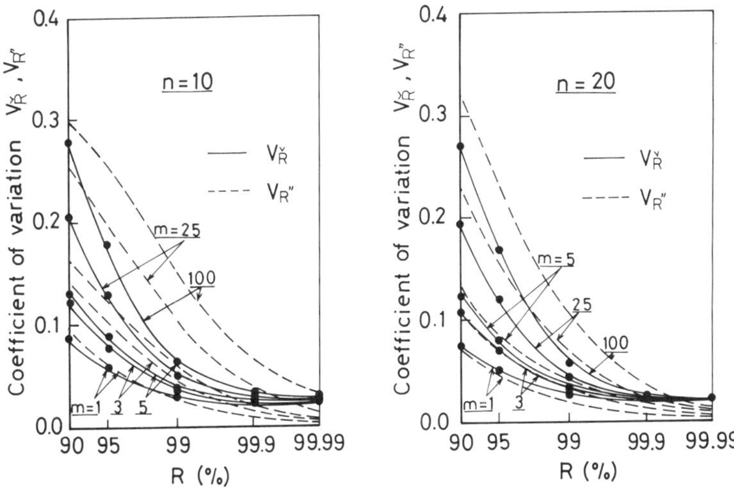

FIG. 18. Coefficients of variation, $V_{\check{R}}$ and $V_{R''}$ as a function of fleet reliability R ($n = 10$ and 20).

CONCLUSION

The concept of the statistical scatter factor has been applied to the case in which the parent distribution of the fatigue life of aircraft or their components is a two-parameter Weibull with both the shape and scale parameters being unknown. Procedures have been established to evaluate the scatter factor in its extended form using the maximum likelihood estimates of the parameters. The fleet reliability could then be estimated on the basis of the scatter factor of extended form thus evaluated. The effect of the sample size to be used in the fatigue test, of the fleet size and of the reliability level on the accuracy of such an estimation has also been discussed.

REFERENCES

[1] M. Shinozuka, Development of reliability-based aircraft safety criteria: an impact analysis, Technical Report AFFDL-TR-76-31 (1976).

[2] A. M. Freudenthal, The scatter factor in the reliability assessment of aircraft structures, *J. Aircraft,* **14**(2), 202 (1977).

[3] H. Ishikawa, T. Tanimoto, H. Kimura and M. Shinozuka, Reliability

assessment of machines and structures with Weibull type of life distribution, *J. Soc. Mater. Sci. Jap.*, **30**(328), 39 (1981).
[4] M. Shinozuka, Reliability-based scatter factors, Technical Report AFFDL-TR-78-17 (1978).
[5] I. C. Whittaker and P. M. Besuner, A reliability analysis approach to fatigue life variability of aircraft structures, Technical Report AFML-TR-69-65 (1969).

Study on the Probabilistic Design of Fiber-Reinforced Composite Material

ZENICHIRO MAEKAWA

Faculty of Textile Science, Kyoto Institute of Technology, Matsugasaki, Sakyoku, Kyoto, Japan

ABSTRACT

A probabilistic analysis method was studied for the strength of fiber-reinforced composite materials under the plane stress state. The proposed method accounts for scatter of the material strength and the applied stress, which is ignored in the conventional design approach. The method applies the first-order-second-moment method to the maximum stress and maximum work criteria, the tensor polynomial theory being used as the maximum work criterion. The fiber-reinforced laminate is treated as a homogeneous and orthotropic material, and the method can be applied to the laminate in on-axis and off-axis conditions. The computational procedure for obtaining the mean value of allowable applied stress is shown in the case of two failure criteria. This procedure is also applied to the stress analysis of a composite laminate with cut-outs by combining the finite element method with the second-moment approximation method.

INTRODUCTION

Fiber-reinforced composite materials are generally characterized by having considerable scatter in the mechanical properties. It is evident that this variability must be taken into account when these composites are used as structural materials. From this point of view, some probabilistic design methods considering the variation of the strength of composites have been proposed [1–3]. However, it is considered necessary to take the variation of applied load into account in the reliability design of composite materials, since the variation of applied load cannot often be ignored in comparison with that of the strength of a composite.

This study deals with a probabilistic design method for fiber-reinforced composite materials. The proposed method accounts for the scatter of material strength and applied load, which are generally ignored in the conventional design approach. The fiber-reinforced laminate is treated as a homogeneous and orthotropic material, and the two failure criteria of maximum stress and maximum work are applied to the composite material under on-axis and off-axis loadings. These criteria are widely used in the fracture theory of fiber-reinforced composites. The probabilistic design method for the strength of the composite material under a plane stress state is established by applying the first-order-second-moment method [4] to the two failure criteria. The proposed method requires only the mean values and the coefficients of variation of both the principal strength components and the applied load. Thus, this method is considered to be simple and effective enough for practical use.

This method can also be applied to the stress analysis of a composite material with an area of stress concentration [5]. It is well known that the finite element procedure is useful for the stress analysis of a notched composite material. Several works concerning stress analysis by the finite element method in practical composites have been reported [6–8]. The method that will be presented involves stress analysis by the finite element method and reliability analysis by the first-order-second-moment method. The stresses in a composite plate with double cut-outs subjected to tensile loading will be analyzed by this probabilistic method.

THEORY

In order to account for the behavior of fiber-reinforced composites under combined stress states, we consider the homogeneous and orthotropic material in a plane stress state under the off-axis loading shown in Fig. 1. When 1 and 2 are the axes of material symmetry, and x and y are the reference coordinate axes of the externally applied stresses, the usual transformation equation in matrix form is

$$\begin{Bmatrix} s_1 \\ s_2 \\ s_6 \end{Bmatrix} = \begin{pmatrix} l^2 & m^2 & -2lm \\ m^2 & l^2 & 2lm \\ lm & -lm & (l^2 - m^2) \end{pmatrix} \begin{Bmatrix} s_x \\ s_y \\ s_s \end{Bmatrix} \tag{1}$$

where $m = \cos\theta$, $l = \sin\theta$ and θ is the angle between the applied

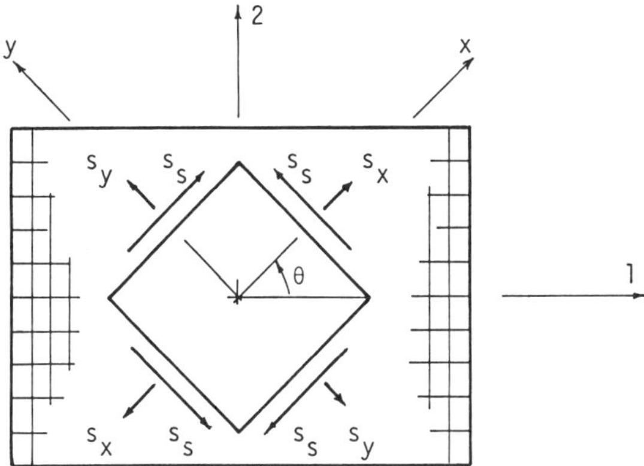

FIG. 1. Relationship between the principal axis and load axis.

stress and the principal axes. On-axis loading can be handled by the off-axis analysis as the special case of $\theta = 0$. s_1, s_2 and s_6 are the stress components with respect to the principal material axes, and s_x, s_y and s_s are those of the load axes. S_1, S_2 and S_6 are strength components on the principal axes as shown in Fig. 2. When s_1 and/or s_2 are compressive, the corresponding compressive strengths are required. For this paper, we chose the maximum stress and maximum work criteria, which are most common failure criteria for composite materials, so that probabilistic design methods for composite materials based on these failure criteria could be proposed.

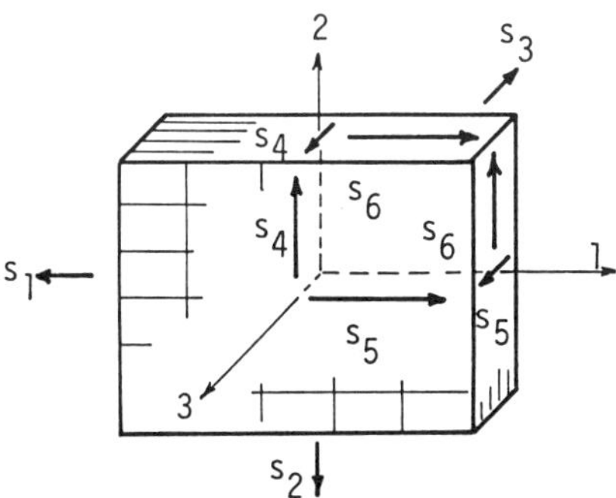

FIG. 2. Explanation of symbols on the principal axis of a laminate and the applied stress.

MAXIMUM STRESS CRITERION

This criterion specifies that failure in a composite material results when one of the applied stresses reaches the limiting value without regard to the other stresses. Therefore, this theory is described by the following inequalities:

$$Z_{m1} = S_1 - s_1 < 0 \qquad Z_{m2} = S_2 - s_2 < 0 \qquad Z_{m6} = S_6 - s_6 < 0 \quad (2)$$

where Z_{mi} $(i = 1, 2, 6)$ denotes the safety margins.

The stress–strength model (S–S model) is now introduced for a probabilistic design approach. The concept of this model is that the stress introduced by operating conditions and the strength of the material are treated as random variables, and if the applied stress exceeds the strength of the material, failure results, as shown in Fig. 3. As S_i $(i = 1, 2, 6)$ denotes the random strength variables and s_i $(i = 1, 2, 6)$ the random stress variables, the safety margins (Z_{mi}) also become random variables, from Eqn. (2).

The probabilities of failure in each principal direction (p_{fi}) are given by

$$p_{fi} = \text{Prob}[Z_{mi} < 0] = G_{Z_{mi}}(0) \quad (i = 1, 2, 6) \qquad (3)$$

where $G_{Z_{mi}}$ shows the distribution functions of Z_{mi}. Using the relationship of $\xi = (z - \mu_Z)/\sigma_Z$, Eqn. (3) may be transformed as follows:

$$p_{fi} = G_{\xi_{mi}}\left(-\frac{\mu_{Z_{mi}}}{\sigma_{Z_{mi}}}\right) \quad (i = 1, 2, 6) \qquad (4)$$

where μ and σ are the mean and the standard deviation of the subscripted values, respectively. The relationship $Q_{mi} = \mu_{Z_{mi}}/\sigma_{Z_{mi}}$ and

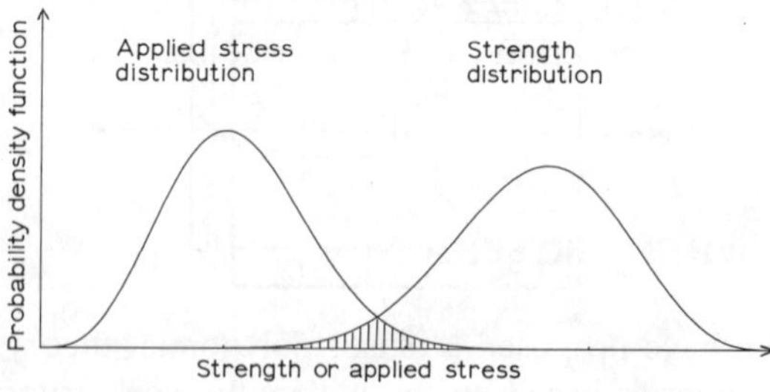

FIG. 3. Explanation of the S–S model.

the use of Eqn. (4) lead to the following equation:

$$Q_{mi} = -G_{\xi_{mi}}^{-1}(p_{fi}) \quad (i = 1, 2, 6) \tag{5}$$

Q_{mi} is called the safety index, and is used as a measure of reliability. When the distribution function $G_{\xi_{mi}}$ is known, Q_{mi} can be obtained from p_{fi}. By using a first-order approximation, the mean and the standard deviation of the safety margins can be calculated as

$$\mu_{Z_{mi}}/\sigma_{Z_{mi}} = (1 - F_i)/\sqrt{C_{S_i}^2 + F_i^2 C_{s_i}^2} \quad (i = 1, 2, 6) \tag{6}$$

where C is the coefficient of variation of the subscripted values. F_i denotes the reciprocal of the central safety factor shown by

$$F_i = \mu_{s_i}/\mu_{S_i} \quad (i = 1, 2, 6) \tag{7}$$

Using the first-order approximation, the transformation equation with respect to the mean values of the applied stresses is given by

$$\begin{Bmatrix} \mu_{s_1} \\ \mu_{s_2} \\ \mu_{s_6} \end{Bmatrix} = \begin{pmatrix} l^2 & m^2 & -2lm \\ m^2 & l^2 & 2lm \\ lm & -lm & (l^2 - m^2) \end{pmatrix} \begin{Bmatrix} \mu_{s_x} \\ \mu_{s_y} \\ \mu_{s_s} \end{Bmatrix} \tag{8}$$

In the same manner as for the mean value of applied stress, the transformation equation for the variation of applied stress is given by

$$\begin{Bmatrix} C_{s_1}^2 \\ C_{s_2}^2 \\ C_{s_6}^2 \end{Bmatrix} = \begin{bmatrix} \dfrac{l^2}{\mu_{s_1}^2} & \dfrac{m^2}{\mu_{s_1}^2} & \dfrac{-lm}{\mu_{s_1}^2} \\ \dfrac{m^2}{\mu_{s_2}^2} & \dfrac{l^2}{\mu_{s_2}^2} & \dfrac{2lm}{\mu_{s_2}^2} \\ \dfrac{lm}{\mu_{s_6}^2} & \dfrac{-lm}{\mu_{s_6}^2} & \dfrac{(l^2 - m^2)}{\mu_{s_6}^2} \end{bmatrix} \begin{Bmatrix} \mu_{s_x}^2 & C_{s_x}^2 \\ \mu_{s_y}^2 & C_{s_y}^2 \\ \mu_{s_s}^2 & C_{s_s}^2 \end{Bmatrix} \tag{9}$$

Thus, the probability of failure for composite materials in the case of the maximum stress criterion can be described as

$$p_f = 1 - (1 - p_{f1})(1 - p_{f2})(1 - p_{f6}) \tag{10}$$

MAXIMUM WORK CRITERION

Tsai and Wu have proposed a tensor polynomial theory, which may cover various theories based on the maximum work criterion [9]. The form of the failure criterion for a composite material under plane stress loading can be described as

$$Z_e = 1 - \left[\sum_{i=1}^{2} \left[s_i \left(\frac{1}{S_i} - \frac{1}{S_i'} \right) + \frac{s_i^2}{S_i S_i'} \right] + \frac{s_6^2}{S_6^2} + \frac{2k_{12} s_1 s_2}{\sqrt{S_1 S_1' S_2 S_2'}} \right] \leq 0 \tag{11}$$

where S_i and S_i' ($i = 1, 2$) are the tensile and compressive strengths on the principal axes, respectively, and k_{12} is an interaction term whose absolute value is less than unity. The failure of the composite material is assumed to occur if Z_e in Eqn. (11) is negative. Substituting Eqn. (1) into Eqn. (11), it follows that Z_e is a function of s_x, s_y, S_s, S_1, S_1', S_2, S_2' and S_6. These variables are assumed to be independent random variables. By expanding Eqn. (11) in a Tayler series about the mean of each variable and using the first-order approximation, the mean value of Z_e is given by

$$\mu_{Z_e} = 1 - \left[\left(1 - \frac{1}{r_{\mu 1}}\right) F_1 + \left(1 - \frac{1}{r_{\mu 2}}\right) F_2 + \frac{1}{r_{\mu 1}} F_1^2 + \frac{1}{r_{\mu 2}} F_2^2 + \frac{2k_{12}}{\sqrt{r_{\mu 1} r_{\mu 2}}} F_1 F_2 \right] \tag{12}$$

Similarly, the variance of Z_e can be shown as

$$\sigma_{Z_e}^2 = (C_{s_x} \zeta_x)^2 + (C_{s_y} \zeta_y)^2 + (C_{s_s} \zeta_s)^2 + C_{S_1}^2 (\zeta_1^2 + r_{C1} \zeta_1'^2) + C_{S_2}^2 (\zeta_2^2 + r_{C2}^2 \zeta_2'^2) + (C_{S_6} \zeta_6)^2 \tag{13}$$

where $r_{\mu i}$ and r_{ci} are described by

$$\left. \begin{aligned} \mu_{S_i}' &= r_{\mu i} \mu_{S_i} \\ C_{S_i}' &= r_{Ci} C_{S_i} \end{aligned} \quad (i = 1, 2, 6) \right\} \tag{14}$$

and

$$\begin{Bmatrix} \zeta_x \\ \zeta_y \\ \zeta_s \end{Bmatrix} = \begin{bmatrix} \frac{\mu_{s_x}}{\mu_{s_1}} l^2 & \frac{\mu_{s_x}}{\mu_{s_2}} m^2 & \frac{\mu_{s_x}}{\mu_{s_6}} lm \\ \frac{\mu_{s_y}}{\mu_{s_1}} m^2 & \frac{\mu_{s_y}}{\mu_{s_2}} l^2 & \frac{\mu_{s_y}}{\mu_{s_6}} lm \\ -\frac{\mu_{s_s}}{\mu_{s_1}} 2lm & \frac{\mu_{s_s}}{\mu_{s_2}} 2lm & \frac{\mu_{s_s}}{\mu_{s_6}} (l^2 - m^2) \end{bmatrix} \begin{Bmatrix} (\zeta_1 - \zeta_1') \\ (\zeta_2 + \zeta_2') \\ \zeta_6 \end{Bmatrix} \tag{15}$$

$$\left.\begin{aligned}
\zeta_1 &= F_1 + \frac{F_1^2}{r_{\mu 1}} + \frac{k_{12}}{\sqrt{r_{\mu 1} r_{\mu 2}}} F_1 F_2 \qquad \zeta_6 = 2F_6^2 \\
\zeta_1' &= \frac{F_1}{r_{\mu 1}} + \frac{F_1^2}{r_{\mu 1}} + \frac{k_{12}}{\sqrt{r_{\mu 1} r_{\mu 2}}} F_1 F_2 \\
\zeta_2 &= F_2 + \frac{F_2^2}{r_{\mu 2}} + \frac{k_{12}}{\sqrt{r_{\mu 1} r_{\mu 2}}} F_1 F_2 \\
\zeta_2' &= \frac{F_2}{r_{\mu 2}} + \frac{F_2^2}{r_{\mu 2}} + \frac{k_{12}}{\sqrt{r_{\mu 1} r_{\mu 2}}} F_1 F_2
\end{aligned}\right\} \qquad (16)$$

For the special case of unidirectional loading of the laminate in the off-axis state, the variation of Z_e can be simplified to

$$\begin{aligned}
\sigma_{z_e}^2 = {} & C_{s_x}^2(\zeta_1 + \zeta_1' + \zeta_2 + \zeta_2' + \zeta_6)^2 + C_{S_1}^2(\zeta_1^2 + r_{c1}^2 \zeta_1'^2) \\
& + C_{S_2}^2(\zeta_2^2 + r_{c2}^2 \zeta_2'^2) + C_{S_6}^2 \zeta_6^2
\end{aligned} \qquad (17)$$

and Z_e can then be reduced as in the case of $\theta = 0$

$$\begin{aligned}
\sigma_{Z_e}^2 = {} & C_{s_x}^2(\zeta_1 + \zeta_1')^2 + C_{s_y}^2(\zeta_2 + \zeta_2')^2 + C_{s_s}^2 \zeta_6^2 + C_{S_1}^2(\zeta_1^2 + r_{c1}^2 \zeta_1'^2) \\
& + C_{S_2}^2(\zeta_2^2 + r_{c2}^2 \zeta_2'^2) + C_{S_6}^2 \zeta_6^2
\end{aligned} \qquad (18)$$

Thus, the probability of failure based on the maximum work criterion can be calculated by

$$p_f = \text{Prob.}[Z_e < 0] = G_{Z_e}(0) = G_{\xi_e}\left(-\frac{\mu_{Z_e}}{\mu_{Z_e}}\right) \qquad (19)$$

If p_{fa} is the allowable probability of failure, the nondestructive condition of composite materials is

$$\frac{\mu_{Z_e}}{\sigma_{Z_e}} > -G_{\xi_e}^{-1}(p_{fa}) = Q_{ea} \qquad (20)$$

where Q_{ea} is the safety index corresponding to the allowable probability of failure.

COMPUTATIONAL PROCEDURE

This study deals with the probabilistic design approach for the strength of a composite laminate and the stress analysis of a notched composite plate by the reliability concept.

Probabilistic Design Method

The structural design of a composite materials may often involve the procedure to determine such dimensions as the cross-sectional area and thickness when the characteristics of the applied load the strength of the material are given. Such a procedure reduces to the problem of determining the allowable stress that can be applied to the component.

First, the flow chart for obtaining the mean value of the allowable applied stress in the case of the maximum stress criterion is shown in Fig. 4. The input data necessary for the present procedure are as follows: (1) the mean values and the coefficients of variation of the material strengths in the principal axes, (2) the coefficients of variation of the applied stresses, (3) the angle between the load axis and the principal axis of the material, and (4) the probability of failure. The values of p_{f1}, p_{f2} and p_{f6} satisfying Eqn. (10) are computed in the range not greater than 0·5, since the values of the safety should be greater than unity. Using Eqns. (5), (6) and (8), we can obtain every set of μ_{s_x}, μ_{s_y} and μ_{s_s} having the given probability of failure.

Secondly, the flow chart for the computational procedure based on the maximum work criterion is shown in Fig. 5. In this case, the input data are the same as in the case of the maximum stress criterion. μ_{s_x}, μ_{s_y} and μ_{s_s}, with initial values of zero, are increased gradually in increments of the small values $\Delta\mu_{s_x}$, $\Delta\mu_{s_y}$ and $\Delta\mu_{s_s}$, respectively. The judgment whether or not the nondestructive condition of Eqn. (20) is satisfied can be made for each set of the mean values of applied stress. Thus, the range of mean values of allowable applied stress is decided under the given probability of failure.

Stress Analysis of a Notched Plate

The proposed method combines stress analysis with the finite element method and reliability analysis with the first-order-second moment method. This method will be applied to a composite plate with double cut-outs subjected to tensile stress.

The flow chart for obtaining the probability of failure in each element is shown in Fig. 6, the input data necessary for the procedure being shown on the flow chart. The composite structure is divided into volume elements having finite dimensions, and the normal stress and shear stress in each element are then computed by the finite element method. The probability of failure in each element is calculated by reliability analysis, where input data of the mean values and coefficients of variation of the applied stress and the material strength

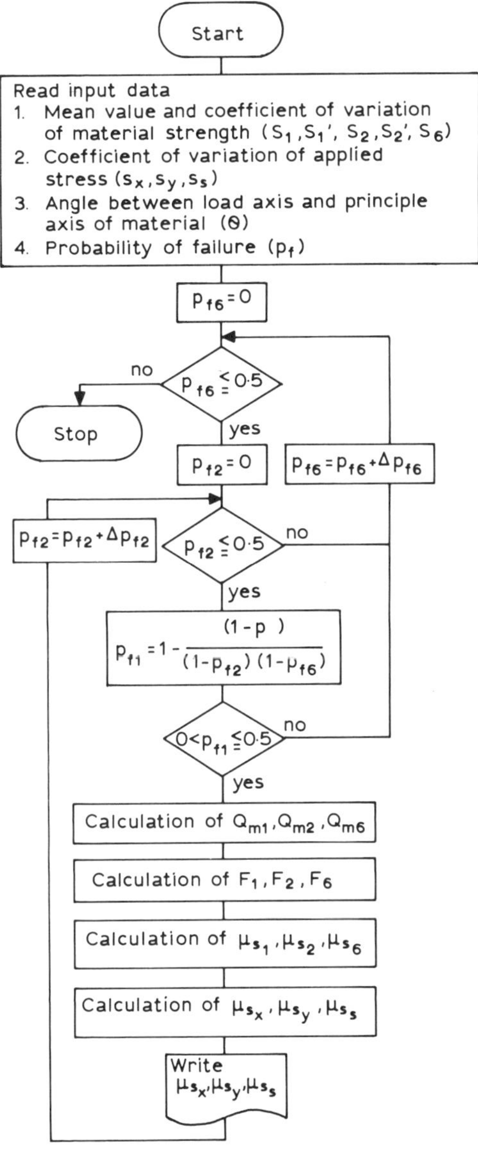

FIG. 4. Flow chart for obtaining the mean values of applied stresses in the case of the maximum stress criterion.

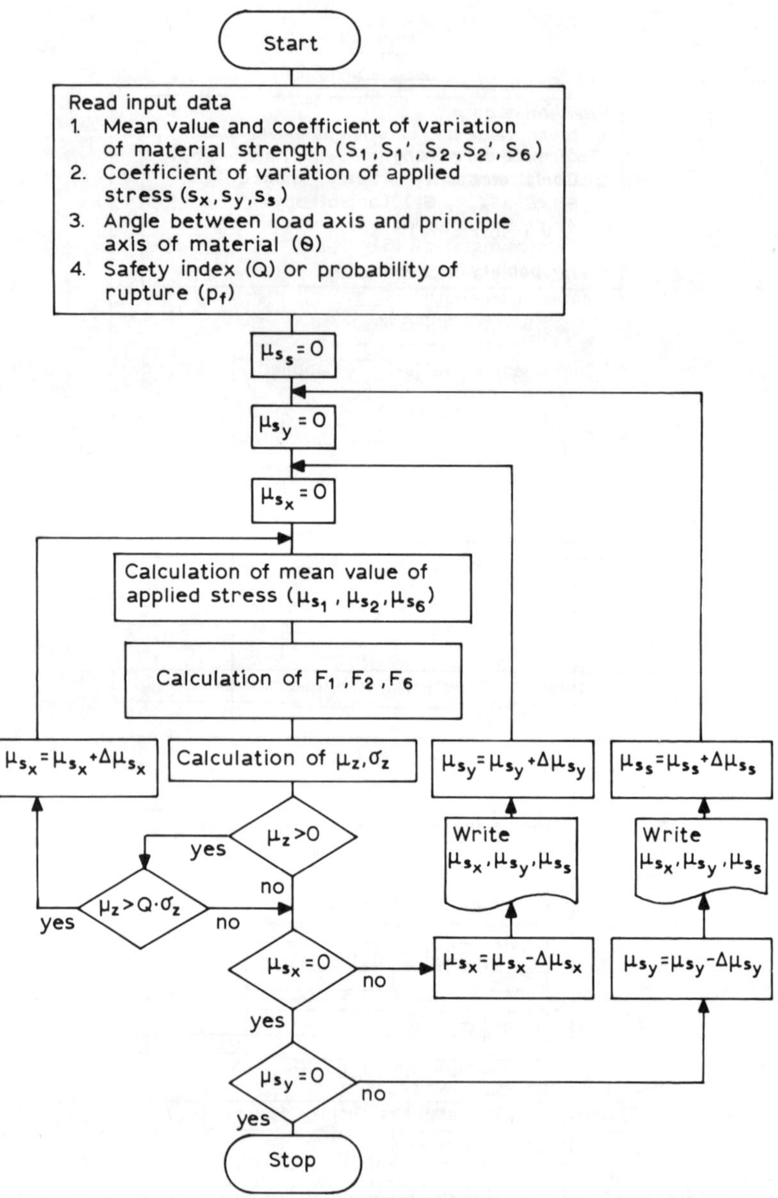

FIG. 5. Flow chart for obtaining the mean values of applied stresses in the case of the maximum work criterion.

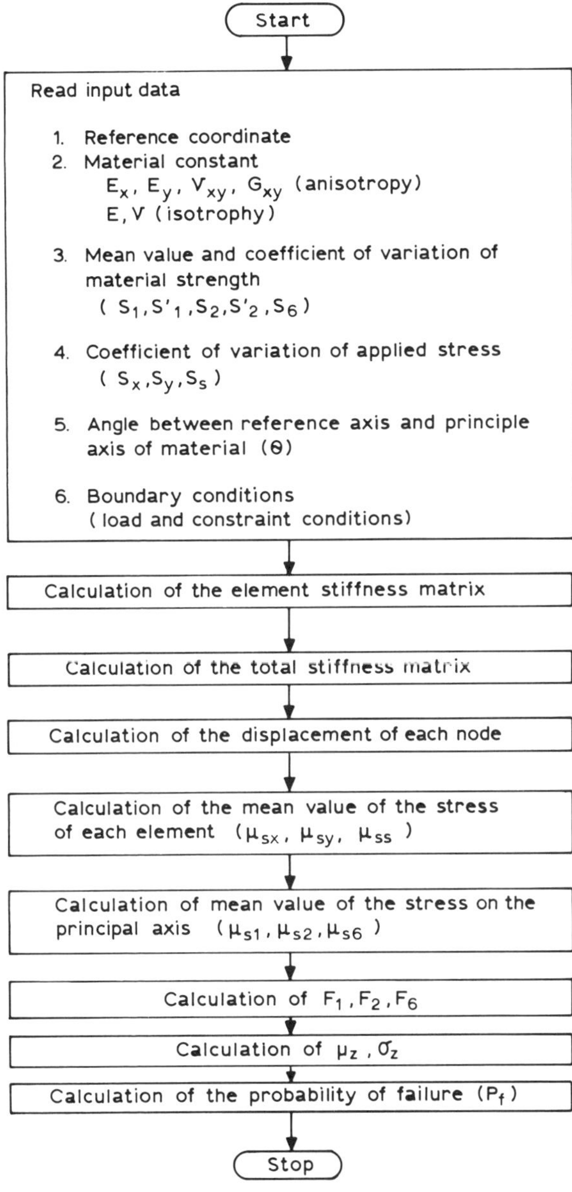

FIG. 6. Flow chart for obtaining the probability of failure in each element of a composite laminate.

are used. The greatest value of the calculated probability of failure in all elements is considered to agree with the probability of failure in composite laminates in the areas of stress concentration. For cases in which the greatest value of the calculated probability of failure is different from the required value, correction to the dimensions of the composite structure and repetition of this computational procedure are carried out.

NUMERICAL EXAMPLES AND CONSIDERATIONS

Probabilistic Design Method

The proposed probabilistic design method was applied to a unidirectional glass fiber-reinforced epoxy laminate. The experimental data for on-axis and off-axis loading used in the present study were equated from the paper by Tsai [10]. The mean values and shear strengths (S_6) of the laminate are illustrated in Table 1, the data concerning the coefficient of variation in this table having been obtained from Jones' data [11].

First, the mean values of the allowable applied stresses were calculated for the laminate under on-axis loading. The numerical results by the computational procedure based on the maximum stress criterion are shown in Fig. 7. The design tolerance region of the on-axis state is shown with the probability of failure and the coefficient

TABLE 1
Data for the unidirectional composite laminate

Strength of laminate	*Mean μ (MPa)*	*Coefficient of variation C(%)*
Tensile strength (S_1)	1032·9	10·0
Compressive strength (S_1')	1032·9	12·0
Tensile strength (S_2)	27·4	11·0
Compressive strength (S_2')	137·2	8·0
Shear strength (S_6)	54·9	6·0

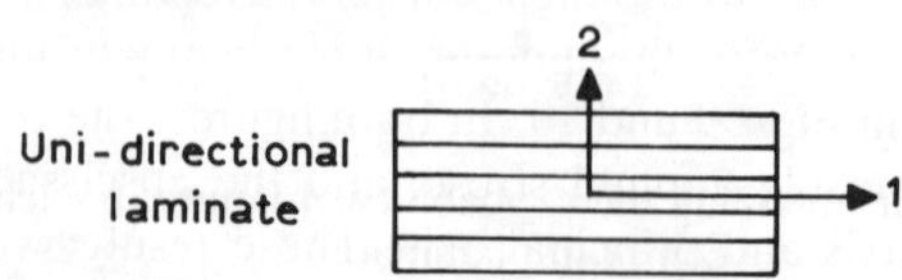

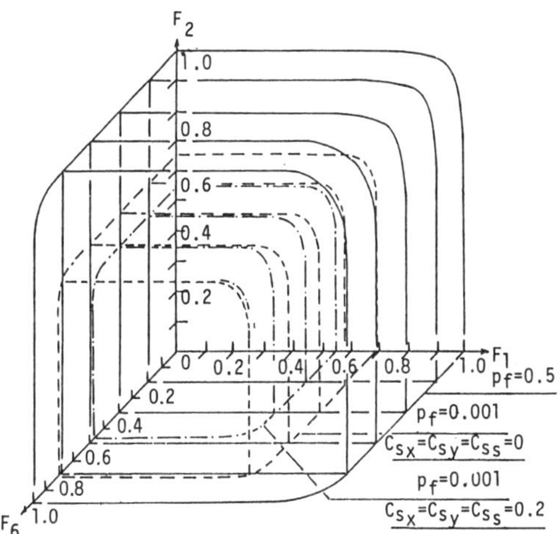

FIG. 7. Tolerance region for the design of a unidirectional laminate in the on-axis condition based on the maximum stress criterion.

of variation of applied stress as parameters. The three reciprocals of the central safety factor on the principal axis F_1, F_2 and F_6 were adopted as the coordinate axes, respectively. In this calculation, the various values represented in Table 1 were used as the strengths of the laminate, and the distribution function of safety margins was assumed to be a normal distribution. It can be seen from this figure that the curved surface of failure has a tendency to decrease as the probability of failure decreases and the coefficient of variation of applied stress increases. It should be noted that the mean values of applied stresses have to be estimated at a smaller value as the scatter of applied stresses increases even if the probability of failure is constant.

Next, the numerical results in the case of the maximum work criterion are illustrated in Fig. 8. Comparing this figure with Fig. 6, it can be seen that the curved surfaces of failure are more conservative than those from the case of the maximum stress theory, since it is generally known that the results obtained by substituting unity into the interaction term k_{12} lead to the most conservative values.

The calculated results for the unidirectional laminate under uniaxial loading are shown in Figs. 9 and 10. In both figures, the ordinate is the mean value of allowable applied stress, and the abscissa is the angle between the load axis and principal axis. These results were obtained

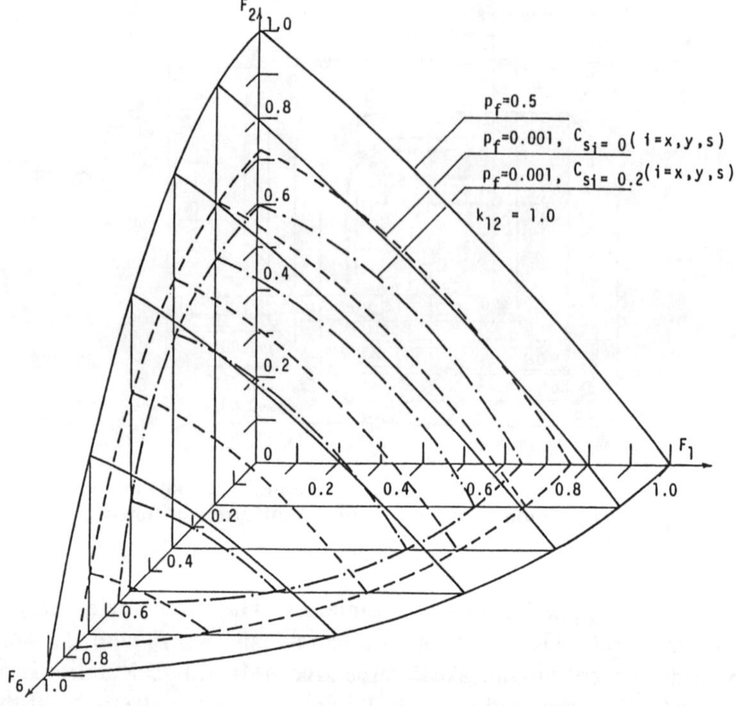

FIG. 8. Tolerance region for the design of a unidirectional laminate in the on-axis condition based on the maximum work criterion.

by the computational procedure based on the maximum work criterion. This calculation was carried out for the special case of $\mu_{s_y} = \mu_{s_s} = 0$ and $C_{s_y} = C_{s_s} = 0$ in the flow chart of Fig. 7, and the variance of Z_e was obtained by using Eqn. (17). Figure 8 shows a comparison of the allowable tensile stress (solid line) and compressive stress (dotted line) with the experimental data of Tsai, where the parameter is the probability of failure. The variance of applied stress (s_x) was treated as zero, since the scatter of the applied stress is not taken into account. It can be seen in this figure that the mean value of the applied stress decreases as the angle on the abscissa increases, and the calculated results are in good agreement with the experimental data for the tensile stress and relatively good for the compressive stress. Figure 9 shows the effect of the coefficient of variation of applied stress (C_{s_x}) on

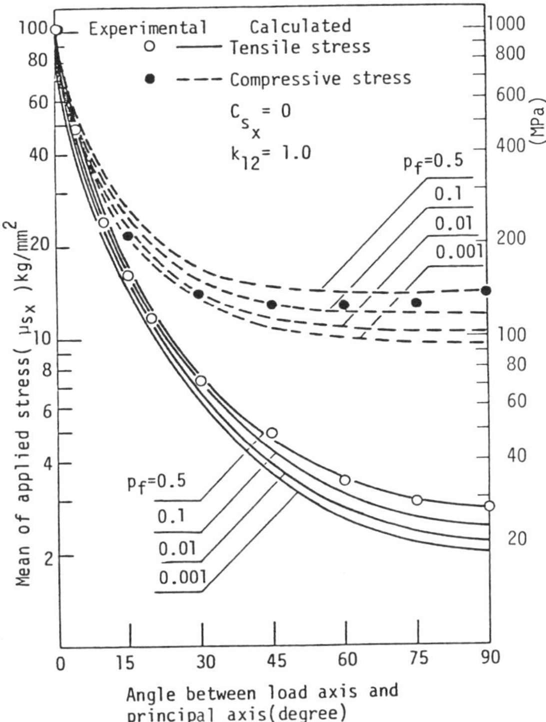

FIG. 9. Relationship between the mean value of allowable applied stress and loading angle with the probability of failure as a parameter in the case of the maximum work criterion.

the mean value of the applied stress (μ_{s_x}), indicating that μ_{s_x} decreases as C_{s_x} increases.

The numerical results for the unidirectional laminate in the on-axis state under uniaxial loading are illustrated in Fig. 11. This figure shows the relationship between F_1 and the safety index with C_{s_x} as a parameter. The data suggest that the central safety factor should be estimated principally as the safety index and the variation of applied stress increase. It should be noted from the various results just described that the mean value of applied stress or the central safety factor must be decided by taking the scatter of the applied stress, the material strength and the given probability of failure into account.

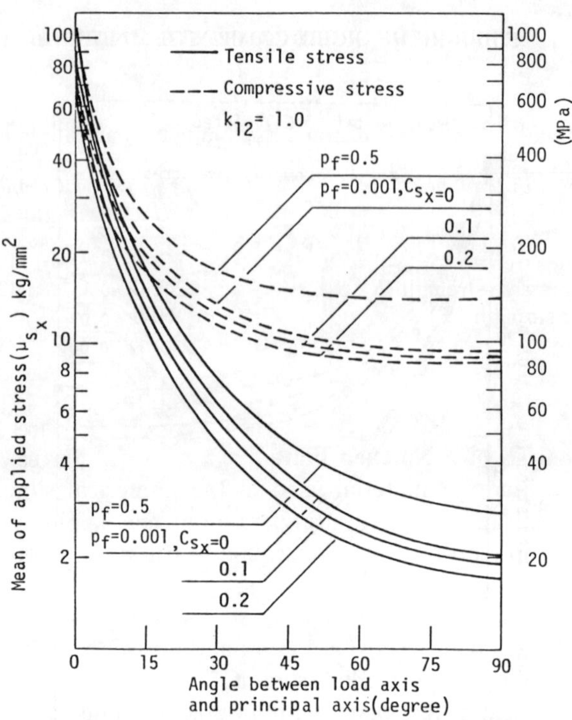

FIG. 10. Relationship between the mean value of allowable applied stress and loading angle with the coefficient of variation of applied stress as a parameter in the case of the maximum work criterion.

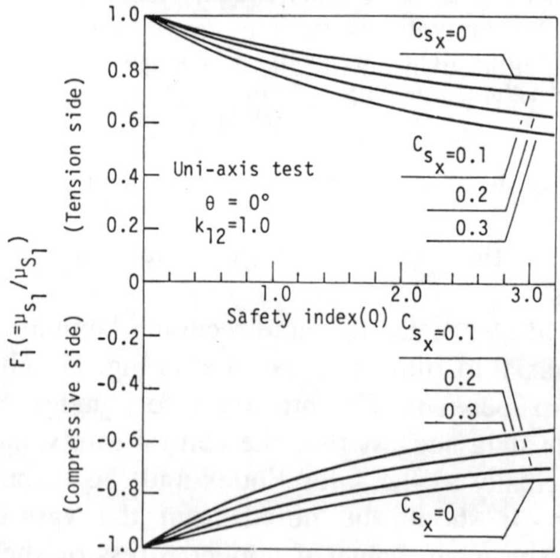

FIG. 11. Relationship between the safety index and F_1 based on the maximum work criterion in the case of only the applied stress s_1 under on-axis loading.

TABLE 2
Data for the chopped strand random mat composite laminate

Strength of laminate	*Mean μ (MPa)*	*Coefficient of variation C(%)*
Tensile strength (S_1, S_2)	112·0	10·0
Compressive strength (S'_1, S'_2)	242·0	12·0
Shear strength (S_6)	60·0	6·0

Stress Analysis of a Notched Plate

The two kinds of material used in the numerical simulation were unidirectional and chopped strand random mat laminates, the latter having isotropic properties. The strength properties of both laminates are shown in Tables 1 and 2. In the case of the unidirectional laminate, the values of Young's modulus (E_1 and E_2) are 125 GPa and 8·5 GPa, the modulus of rigidity (G_{12}) is 5 GPa, and Poisson's ratio $(\nu)_{12}$ is 0·33. In the case of the isotropic laminate, E is 11·25 GPa and ν is 0·33.

Let us consider the plate with double cut-outs subjected to an on-axis tensile stress as shown in Fig. 12. Because of symmetry, the calculation was carried out in only one quadrant of the plate. Along the symmetry axes (OB and OA) the displacements in the X and Y directions are respectively fixed. The probability of failure at each element was calculated in accordance with the procedure of the flow chart in Fig. 6. Part of the calculated results are shown in Fig. 12, in which the coefficient of variance of the applied stress was treated as a parameter, and the symbols classified according to the range of

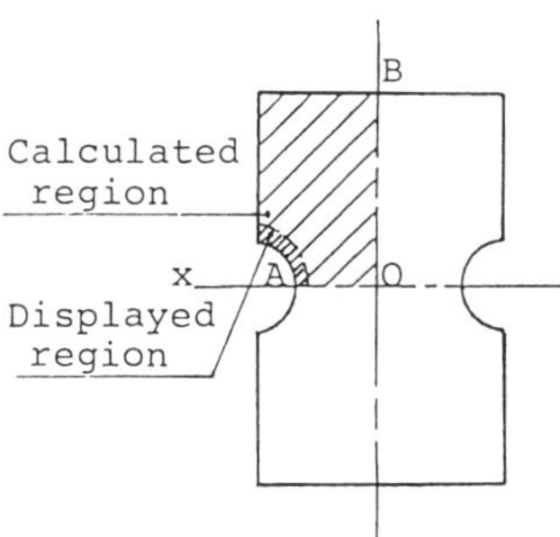

FIG. 12. Finite element model in one quadrant of a composite plate with double cut-outs.

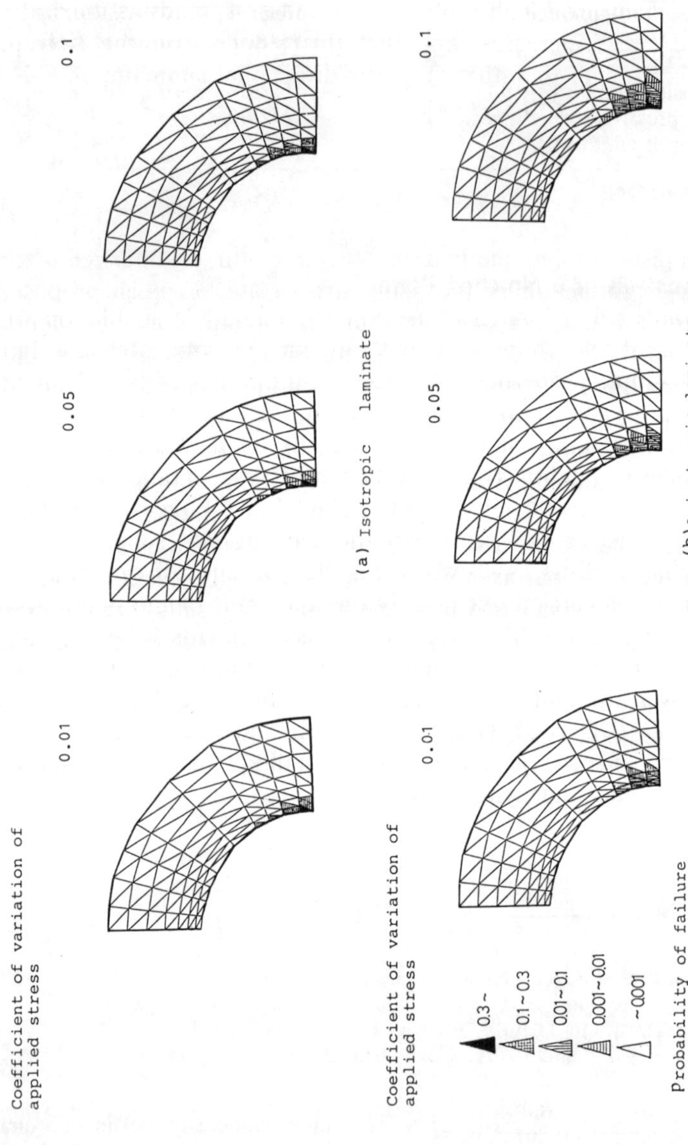

FIG. 13. Distribution of the probability of failure in each element around the cut-out. (a) Isotropic laminate. (b) Anisotropic laminate.

probability of failure. The values for the probability of failure in the elements shown in Fig. 13 are very small. It can be seen in this figure that the domain for a high probability of failure spreads as the scatter of applied stress increases, and that this tendency in the isotropic laminate is stronger than that of the unidirectional laminate.

CONCLUSIONS

A probabilistic design method for the strength of fiber-reinforced composite materials under the plane stress state has been proposed. This method is characterized by both the scatter of the material strength and the applied stress being considered. This design method was accomplished by applying the second-moment approximation method to the maximum stress criterion and the maximum work criterion. Furthermore, in order to obtain the probability of failure in a composite laminate with stress concentration, the finite element method was included in this analysis. The computational procedures for obtaining the mean value of the applied stress or the central safety factor on the principal axes were described for the cases of the two criteria. It can be concluded from the numerical results that the mean values of applied stresses or the central safety factors on the uniaxial axes have to be decided by taking the scatter of the applied stress, the material strength and the probability of failure into account. The proposed method is considered to be effective for practical use, since it requires only the mean values and the variations of both the material strength and the applied stress to be used.

REFERENCES

[1] C. T. Sun and S. E. Yamada, *J. Compos. Mater.*, **12,** 169 (1978).
[2] S. E. Yamada and C. T. Sun, *J. Compos. Mater.*, **12,** 275 (1978).
[3] Z. Maekawa and T. Fujii, *Proc. ICCM-IV,* 537 (1982).
[4] A. H. S. Ang and C. A. Cornell, *J. Struct. Div., Proc. ASCE,* 1755 (1974).
[5] Z. Maekawa, H. Hamada, T. Horino and H. Nakayasu, *Proc. Int. Symp. Composites '86* (1986) in press.
[6] Z. Maekawa, A. Kaji, H. Hamada and M. Nagamori, *Proc. ICCM-V,* 99 (1985).

[7] E. F. Rybicki and D. W. Schmueser, *J. Compos. Mater.*, **12,** 300 (1978).
[8] C. M. S. Wong and F. L. Matthews, *J. Compos. Mater.*, **15,** 481 (1981).
[9] S. W. Tsai and E. M. Wu, *J. Compos. Mater.*, **5,** 58 (1971).
[10] S. W. Tsai, *Fundamental Aspects of Fiber Reinforced Plastic Composites,* Wiley, New York, 3 (1968).
[11] B. H. Jones, *ASTM STP 460,* 307 (1969).

Strength of Alumina Against Static and Impact Loads

TOSHIRO YAMADA and JUNICHI KITAZUMI

Department of Mechanical Engineering, Niihama National College of Technology, Niihama city, Ehime, Japan

ABSTRACT

Three-point bending tests and the impact tests were performed on three kinds of alumina ceramics with different impurities, and the fracture toughness of these materials was investigated by the indentation microfracture method.

In the constant loading rate bending test, the distribution of failure probability of the three kinds of alumina differed according to the influence of impurities.

The distribution of impact values was investigated experimentally, and compared with the distribution of static strength. The effect of impurities on the flexural strength was greater than on the impact values and fracture toughness.

INTRODUCTION

It is well known that the variation of the strength of ceramics under various conditions is very large, and that the weak point of ceramics is their unique brittleness when compared with metallic materials. In the application of ceramic materials to machine structures, the reliability of the materials should be guaranteed against various loadings that are expected to be applied in service. Many structures are commonly subjected to dynamic loads and are sometimes damaged by such loads. The brittleness of ceramics often limits their use as structural materials, particularly under conditions of impact loading.

The brittleness of ceramics is generally estimated by the stress intensity factor. However, the stresses produced by impact loading are different from those arising from ordinary static loads. For example, the impact stresses are influenced by the modulus of elasticity, the mass, the shape, the velocity of propagation of the stress wave and the particle velocity in the material. It is, therefore, inadequate to estimate the impact strength only by the stress intensity factor or the fracture toughness (K_{IC}). The strength of a material against impact needs to be obtained directly from experiments for the various impacts (bending, tension, compression and torsion).

A little research on the absorbed energy of ceramics by an instrumented impact test has been reported, [1, 2] but no authorized method for impact tests on ceramics exists at present.

In this study, the impact strengths of three kinds of alumina with different impurities were analyzed by Izod impact tests, and the cumulative probabilities of the impact value were compared with the probabilities of the three-point bending strength. The effects of the impurities in alumina on these strengths will be discussed on the basis of the experimental results.

The relationship between the impact strength for bending and the fracture toughness will also be discussed, and the fracture toughness that is desirable for ceramics (alumina) when applied to components subjected to dynamic loads will be proposed. It is convenient to estimate the brittleness of ceramics by their impact value as well as by the stress intensity factor for machine design so that we can compare the impact strength of ceramics with that of metallic materials.

EXPERIMENTAL PROCEDURE

Table 1 shows the chemical composition of the alumina ceramics employed here. Test pieces of these alumina ceramics had been fired at 1600 °C, and were shaped and finished to rectangular-section bars with dimensions of 3 mm × 8·5 mm × 54 mm by polishing their surfaces with diamond paste of #1500. These three kinds of alumina are

TABLE 1
Chemical composition of the tested alumina materials (wt. %)

Material	*Al_2O_3*	*SiO_2*	*CaO*	*MgO*
Al_2O_3-1	95·38	3·08	1·02	0·51
Al_2O_3-2	94·97	3·59	0·21	1·95
Al_2O_3-3	99·37	0·38	0·12	0·01

designated Al_2O_3-1, Al_2O_3-2 and Al_2O_3-3 (Table 1), with respective densities of 3·59, 3·72 and 3·83 g cm^{-3}.

The three-point bending tests under a constant loading rate were carried out in an ambient atmosphere on an Instron universal testing machine. In these bending tests, the span length was 30 mm [3], and three steel columns of 5 mm diameter were used as the loading and supporting points on the specimens. The loading and supporting blocks were supported on three steel balls of 10 mm diameter so as not to cause any torsional load on the specimens. The stress rate was controlled at 100 MPa s^{-1} at the point subjected to the maximum nominal stress.

The Izod impact tests were carried out in an ambient atmosphere on a Charpy impact tester (0·98 N-m capacity) with a vice which had been specially made to secure one end of the specimen. The cross-section of each specimen was rectangular (3 mm in depth and 8·5 mm in width) and was without notches. The length between the vice securing the specimens and the striking edge of the pendulum was 22 mm.

The energy absorbed in breaking the specimen for metallic materials is usually equal to the difference between the energy in the pendulum before and after impact. However, since the energy absorbed by ceramics is very small when compared to metallic materials, a correction for the loss of energy due to air resistance, friction of the machine bearings and the trajectory of the broken sample must be considered to obtain accurate results.

The energy to fracture a specimen may be computed as follows:

$$E_t = E_0 - E_1 - e_f - e_v \tag{1}$$

where E_t is the energy to fracture a specimen, E_0 is the initial potential energy of the pendulum, E_1 the potential energy of the pendulum after fracture, e_f the energy losses due to bearing friction in the machine and air resistance to the pendulum movement and e_v the energy for the trajectory of the broken specimen (it was assumed that the broken specimen leaves the vice with the same speed of 2·92 m s^{-1} as the swinging pendulum).

Supposing that the impact value of a ceramic material is equal to the energy per unit area of cross-section to fracture the specimen, the impact value corresponds to the Charpy impact value of metallic materials.

The fracture toughness was investigated by an indentation micro-fracture method, and the values were calculated by the following equation which was proposed by Anstis *et al.* [4]:

$$K_{IC} = 0{\cdot}016(E/H)^{0{\cdot}5}Wc^{-1{\cdot}5} \quad (2)$$

where K_{IC} is the fracture toughness, W the indentation load, $2c$ the crack length, E Young's modulus and H the hardness.

The distributions of impurities and flaws in the specimens were observed at the fracture section through an X-ray micro-analyzer.

EXPERIMENTAL RESULTS AND DISCUSSION

Figures 1–3 present the relationships between the flexural fracture stress (σ) and the cumulative failure probability (P) for Al_2O_3-1, -2 and -3, respectively. The relationships were approximated by a Weibull distribution with two parameters;

$$\left.\begin{aligned} P &= 1 - \exp[-(\sigma/328)^{7{\cdot}70}] \quad \text{for} \quad Al_2O_3\text{-}1 \\ P &= 1 - \exp[-(\sigma/380)^{6{\cdot}75}] \quad \text{for} \quad Al_2O_3\text{-}2 \\ P &= 1 - \exp[-(\sigma/404)^{6{\cdot}04}] \quad \text{for} \quad Al_2O_3\text{-}3 \end{aligned}\right\} \quad (3)$$

The mean flexural strengths of Al_2O_3-1, -2 and -3 by a Weibull

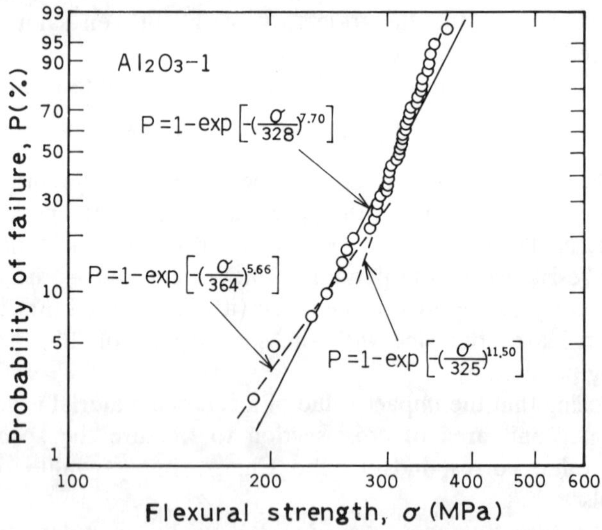

FIG. 1. Weibull distribution of flexural strength for Al_2O_3-1.

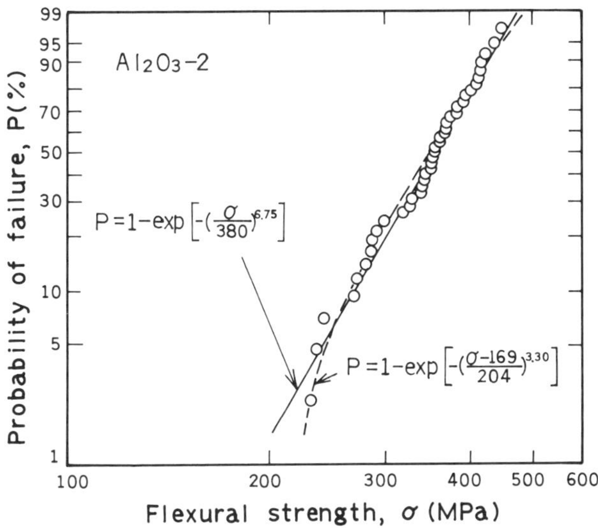

Fig. 2. Weibull distribution of flexural strength for Al_2O_3-2.

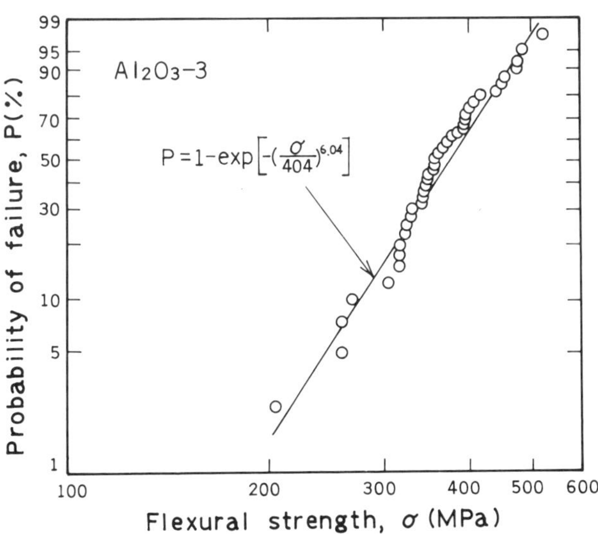

Fig. 3. Weibull distribution of flexural strength for Al_2O_3-3.

distribution with two parameters were 308, 355 and 375 MPa, respectively. For Al_2O_3-1, the cumulative probability of failure can be shown by a multi-modal Weibull distribution. If the probability of failure is divided into two regions ($\sigma > 300$ MPa and $\sigma < 300$ MPa), these are shown by the two dotted lines in Fig. 1.

For Al_2O_3-2, the probability of failure by a Weibull distribution with three parameters is shown by the dotted line in Fig. 2, and can be described by

$$P = 1 - \exp[-\{(\sigma - 169)/204\}^{3 \cdot 30}] \tag{4}$$

The minimum flexural strength of Al_2O_3-2 could be estimated to be 169 MPa, and it was concluded that the values of this expression with three parameters were in good agreement with the experimental results, in comparison with the values for the distribution with two parameters.

Figures 4–6 show the experimental results from the impact tests. The relationships between the impact value (S) and the cumulative probability (P) for Al_2O_3-1, -2 and -3 were approximated by a Weibull distribution with two parameters;

$$\left.\begin{array}{ll} P = 1 - \exp[-(S/5 \cdot 78)^{4 \cdot 64}] & \text{for } Al_2O_3\text{-1} \\ P - 1 - \exp[-(S/5 \cdot 32)^{6 \cdot 47}] & \text{for } Al_2O_3\text{-2} \\ P = 1 - \exp[-(S/5 \cdot 56)^{7 \cdot 27}] & \text{for } Al_2O_3\text{-3} \end{array}\right\} \tag{5}$$

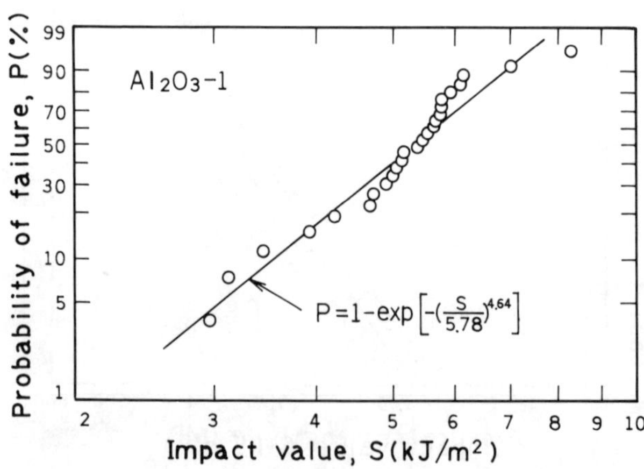

FIG. 4. Weibull distribution of impact values for Al_2O_3-1.

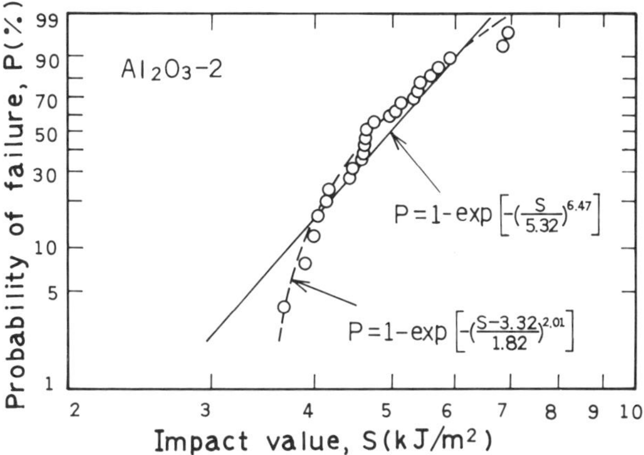

FIG. 5. Weibull distribution of impact values for Al_2O_3-2.

The mean impact values of Al_2O_3-1, -2 and -3, which were estimated by the Weibull distribution with two parameters were 5·3, 5·0 and 5·2 kJ m^{-2}, respectively. For Al_2O_3-2, the distribution of impact values by a Weibull distribution with three parameters is shown by the dotted curved line in Fig. 5, and can be described by

$$P = 1 - \exp[-\{(S - 3{\cdot}32)/1{\cdot}82\}^{2{\cdot}01}] \tag{6}$$

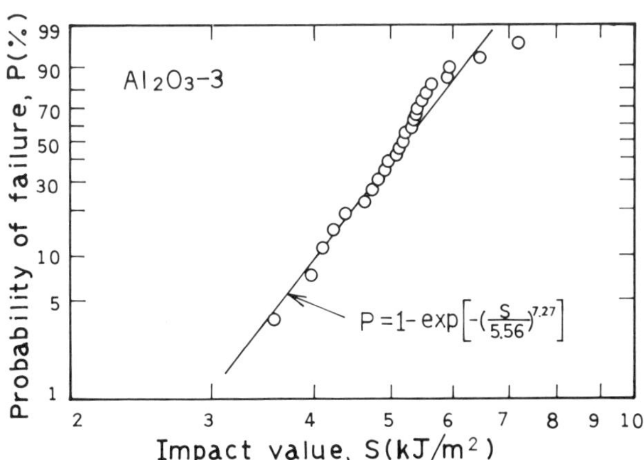

FIG. 6. Weibull distribution of impact values for Al_2O_3-3.

The minimum impact value of Al_2O_3-2 could be estimated to be 3·32 kJ m^{-2}. The three-parameter values of this expression were in good agreement with the experimental impact values for Al_2O_3-2.

Figures 7–9 are X-ray micrographs showing the microstructures and distribution of impurities on the fracture surface produced by the bending tests. Supposing that the pore is shaped like an ellipse, the

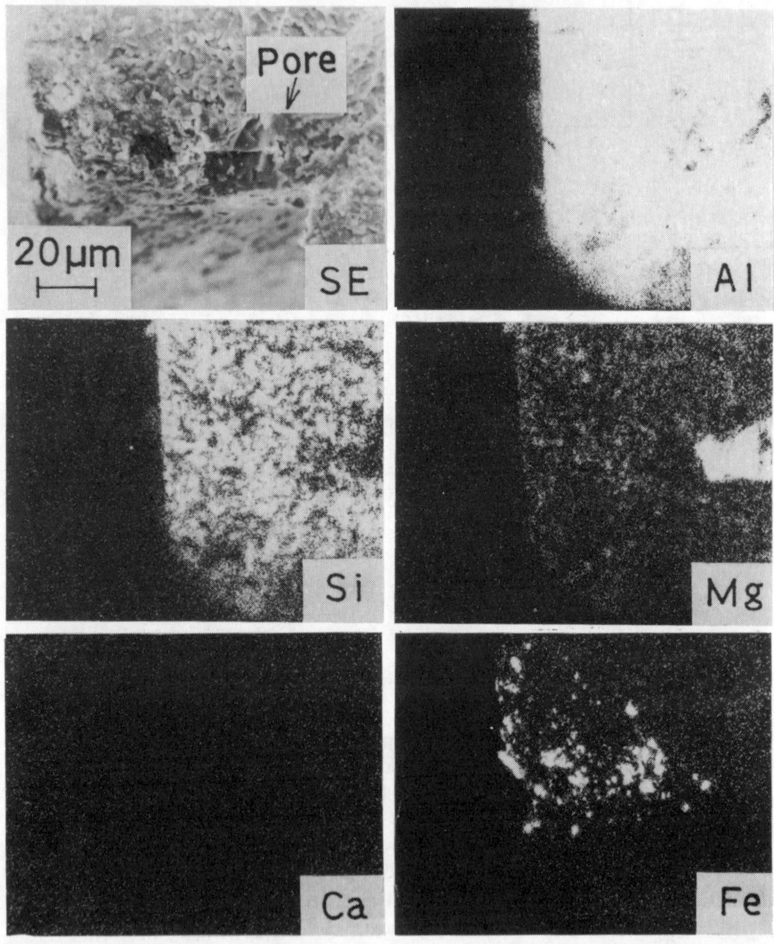

FIG. 7. X-Ray micrographs showing microstructures and the distribution of impurities on the fracture section produced by the bending test on Al_2O_3-1. (SE: secondary electron image.)

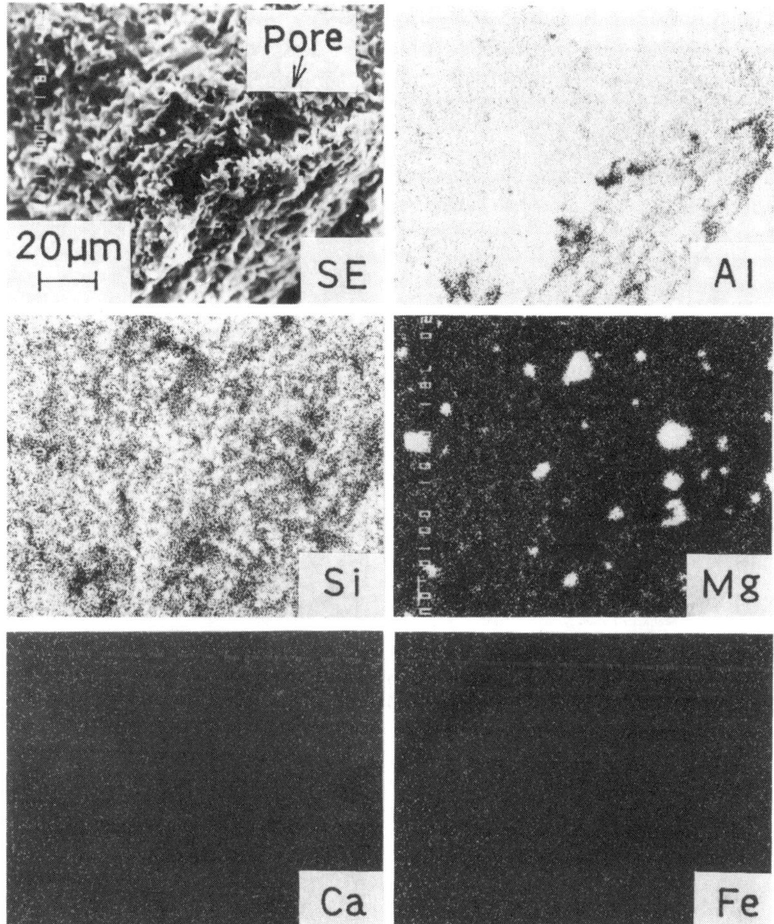

FIG. 8. X-Ray micrographs showing microstructures and the distribution of impurities on the fracture section produced by the bending test on Al_2O_3-2. (SE: secondary electron image.)

large pore size of Al_2O_3-1 was about 55 μm (the long axis) $\times$ 30 μm (the short axis) in Fig. 7 SE, and the pore size of Al_2O_3-2 was about 55 μm $\times$ 20 μm in Fig. 8 SE. For Al_2O_3-3, the pores seen in Al_2O_3-1 and -2 could not be observed. The segregation size of magnesium (Mg) in Al_2O_3-1 was larger than the sizes in Al_2O_3-2 and -3, and the segregation of iron (Fe) was only observed in Al_2O_3-1. Silicon (Si) and

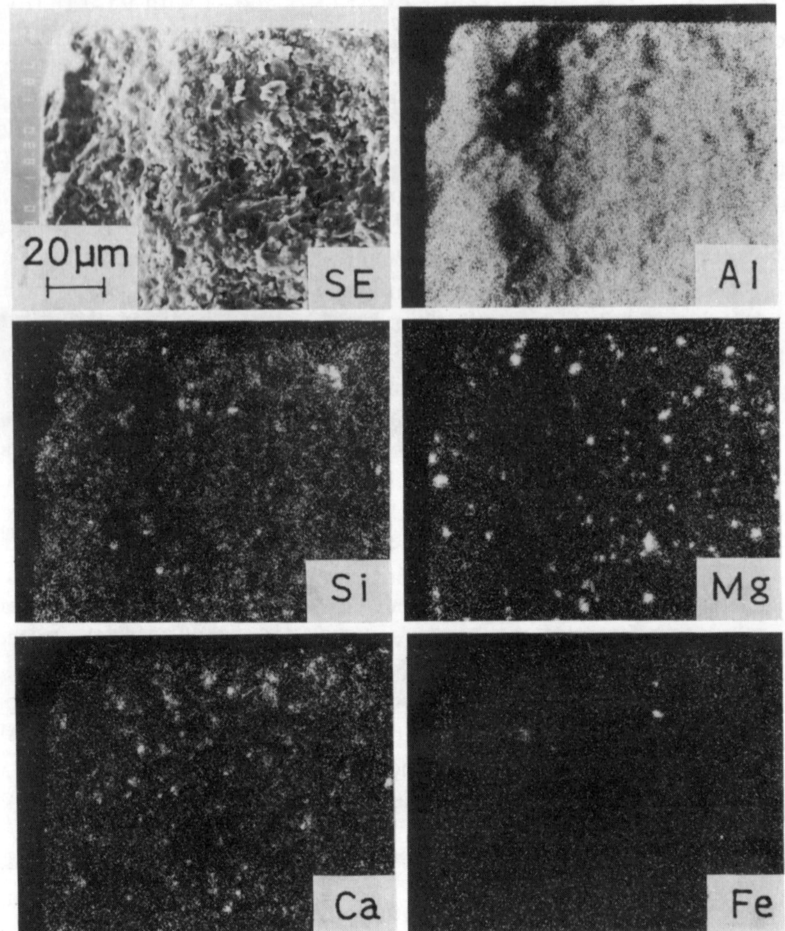

FIG. 9. X-Ray micrographs showing microstructures and the distribution of impurities on the fracture section produced by the bending test on Al_2O_3-3. (SE: secondary electron image.)

calcium (Ca) were uniformly dispersed in these alumina materials, and the large pores of Al_2O_3-1 particularly existed in regions in which the segregation of Mg or Fe existed. Therefore, it is considered that the strength of Al_2O_3-1 with large segregation is statistically lower than the values for the others.

The mean flexural strength of Al_2O_3-3 with 99% purity was 1·21 and

1·06 times as large as the values of Al_2O_3-1 and -2 with 95% purity, respectively. The ratio of the mean flexural strength of Al_2O_3-2 to the value for Al_2O_3-1 was 1·15. This may have come mainly from differences in the segregation of impurities and remaining pores with a different shape, size and number. From the sizes of pores and the density of these alumina materials, it is considered that Al_2O_3-1, -2 and -3 are numbered in the order of their porosity. It has been shown that a larger porosity gives a smaller Young's modulus [5]. The values of Young's modulus for Al_2O_3-1, -2 and -3 that were obtained from the four-point bending tests using strain gages were 280, 320 and 350 GPa, respectively. A linear relationship was observed between the mean flexural strength and Young's modulus for the tested alumina, the ratio of the former to the latter being about $1{\cdot}1 \times 10^{-3}$.

The impact value is a different property of strength from the static strength, and the former does not always correspond to the latter. The effects of impurities on the mean impact values of the tested alumina were small in these tests, but the distribution of the cumulative probability of impact values was similar to the distribution of static strength for these three kinds of alumina. Ceramics for machine components are generally very weak against impact loads when compared with metallic materials, while the impact values of the tested alumina were less than one-hundredth that of regular steels.

Figure 10 shows the relationship between the fracture toughness by the indentation microfracture method and the indentation load. All the plotted points represent the mean of 10 measured values. Assuming that the value at an indentation load of 98 N is the value of

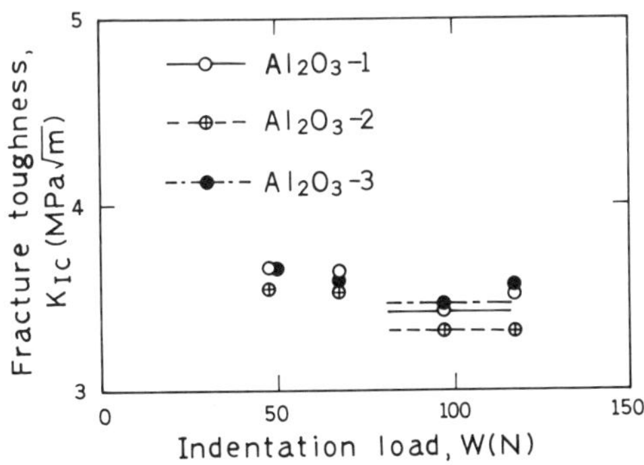

FIG. 10. Relationship between fracture toughness and indentation loads.

TABLE 2
Summary of the mean values of experimental results

Material	*Young's modulus E (GPa)*	*Flexural strength σ (MPa)*	*Impact value S (kJ m^{-2})*	*Fracture toughness K_{IC} (MPa$\sqrt{m}$)*
Al_2O_3-1	280	308	5·3	3·4
Al_2O_3-2	320	355	5·0	3·3
Al_2O_3-3	350	375	5·2	3·4

fracture toughness, the fracture toughness values of Al_2O_3-1, -2 and -3 are 3·4, 3·3 and 3·4 MPa $\sqrt{m}$, respectively, indicating very little difference.

Table 2 shows a summary of the mean values of the experimental results. The relationship between the mean impact value (S) and the fracture toughness (K_{IC}) can be expressed by the following equation:

$$S = A \cdot K_{IC}^2 \tag{7}$$

The value of the proportionality constant (A) in this equation was 0·45 for the aluminas tested.

The mean impact value of the aluminas tested was about 5·2 kJ m^{-2}, and the noted value of regular cast iron is about 100 kJ m^{-2}. Therefore, in order to make use of ceramics instead of cast iron for machine components, it is necessary to increase the toughness of ceramics by improving the processing method or by other techniques. Setting the impact value of alumina equal to the value for regular cast iron in Eqn. (7), we find that the value of K_{IC} is 15 MPa $\sqrt{m}$ for the required value of fracture toughness for alumina. Therefore, it is recommended that the fracture toughness (K_{IC}) for alumina will need to be increased to a value of 15 MPa $\sqrt{m}$ for application to machine components.

CONCLUSION

The strengths against static and impact loads were investigated experimentally for three kinds of alumina with varying impurities.

The cumulative probabilities of static strength have been demonstrated by a Weibull distribution with two parameters, three parameters and a multi-modal type. It cannot be said that a ceramic material that complies with one of these three Weibull distributions is better than ceramics which comply with the others. However, it is certain that a greater homogeneity of inclusions and flaws produces ceramics that can be shown by a Weibull distribution with three parameters, and this easily enables us to have a finite allowable design stress.

The cumulative probabilities of impact values for these alumina materials were similar to the probabilities of static strength for the same materials. The mean impact value of the tested alumina was $5{\cdot}2\ \mathrm{kJ\ m^{-2}}$.

It is concluded that the value for the fracture toughness of alumina which has the same strength against impact loads as regular cast iron, is about $15\ \mathrm{MPa}\sqrt{\mathrm{m}}$. This is a very high value compared with the values for ceramics that are produced at present.

To use ceramic materials for machine components, it will be necessary to create ceramics that have high toughness by improving their processing or by other techniques.

REFERENCES

[1] R. L. Bertolotti, *J. Am. Ceram. Soc.*, **57**(7), 300 (1974).

[2] T. Kobayashi, M. Niinomi, Y. Koide and K. Matsunuma, *J. Jap. Inst. Metals*, **50**(2), 229 (1986).

[3] Testing Method for Flexural Strength (Modulus of Rupture) of High Performance Ceramics, JIS R 1601–1981.

[4] G. R. Anstis, P. Chantikul, B. R. Lawn and D. B. Marshall, *J. Am. Ceram. Soc.*, **64,** 533 (1981).

[5] R. M. Spriggs, *J. Am. Ceram. Soc.*, **45,** 454 (1962).

Recent Developments in the Identification of Dominant Failure Modes and Reliability Assessment for Large-Scale Frame Structures

YOSHISADA MUROTSU[a] and HIROO OKADA[b]

[a] *Department of Aeronautical Engineering,* [b] *Department of Naval Architecture, University of Osaka Prefecture, Sakai, Osaka, Japan*

ABSTRACT

Recent developments in the identification of probabilistically dominant failure modes and in the reliability assessment of large-scale frame structures are presented, based on ultimate collapse analysis. The linearized failure condition of a section is first introduced, which takes account of the combined loading effects of bending moment, axial force and shearing force. This failure criterion facilitates the generation of safety margins and the calculation of failure probabilities. Structural failure is defined as the generation of large deflection due to collapse. The so-called branch-and-bound method combined with the heuristic procedure is then applied to select the probabilistically dominant failure modes, which save the computation necessary to perform a reliability analysis on large-scale structures. Finally, the proposed methods are applied to a transmission line tower and to the transverse ring of a large tanker hull under some notional loading conditions. Through numerical examples, the probabilistic properties of the ultimate collapse of such large-scale frame structures are investigated.

INTRODUCTION

Many studies have been made on the reliability analysis of frame structures, as reviewed by Ditlevsen and Bjerager [1]. However, much work remains to be done for large structures, which have too many failure modes to identify all of them for estimating the system reliability based on ultimate collapse analysis [2–12].

This paper presents recent developments in the reliability assessment of large-scale frame structures based on ultimate collapse analysis. Ultimate collapse is evaluated by using a linearized failure condition of a section under the combined effect of bending moment, shearing force and axial force to generate the safety margins by a matrix method. Probabilistically dominant collapse modes are selected by applying the so-called branch-and-bound method combined with heuristic procedures. These methods are then applied to a transmission line tower, in which the bending moment and axial force dominate the failure criterion, and to the transverse ring of a large tanker hull in which the combined effect of the bending moment, shearing force and axial force determines the plasticity condition. Through numerical examples, the probabilistic properties of the ultimate collapse of these large-scale frame structures are investigated.

GENERATION OF THE STRUCTURAL FAILURE MODES FOR A PLANE FRAME STRUCTURE UNDER THE COMBINED EFFECT OF BENDING MOMENT, SHEARING FORCE AND AXIAL FORCE

Consider a frame structure whose elements are uniform and homogeneous, and to which only concentrated loads and moments are applied. In such a frame structure, the critical sections in which plastic nodes may form are the joints of the elements and the points at which the concentrated loads are applied. The following description is concerned with the case in which various failures occur under the combined loading effects created by bending moment, shearing force and axial force. In the case of plastic collapse, the behavior of members is approximated and structural analysis is performed by combining a plastic node method and a matrix analysis based on the displacement method [9, 13–18].

Derivation of Reduced Stiffness Matrixes and Equivalent Nodal Forces

Let $\mathbf{X}_t = (F_{xi}, F_{yi}, M_{zi}, F_{xj}, F_{yj}, M_{zj})^T$ and $\delta_t = (v_{xi}, v_{yi}, \theta_{zi}, v_{xj}, v_{yj}, \theta_{zj})^T$ denote the nodal force and displacement vectors of the unit element i, j; e.g. the element number t in the local coordinate system shown in Fig. 1(a).

When the interaction of bending moment, shearing force and axial force is considered, the yielding condition of such a structural element as the deep girder comprising the transverse ring of a tanker hull is

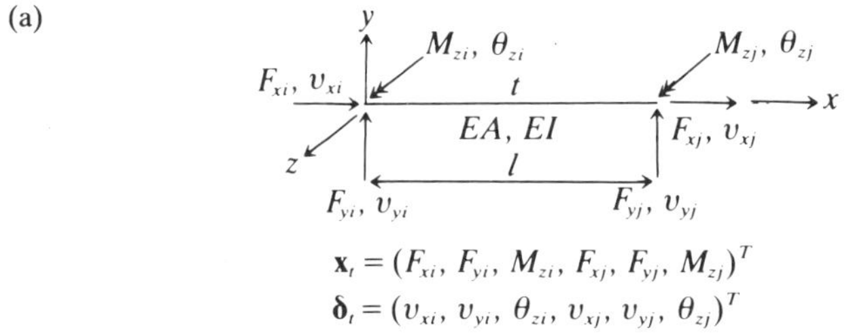

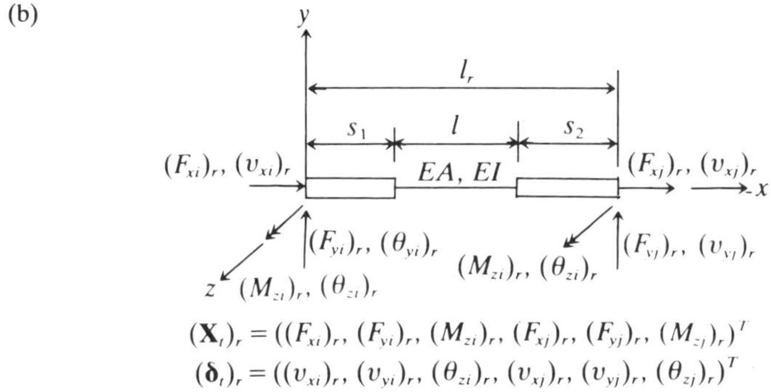

FIG. 1. (a) Nodal forces and displacements of an elasto-plastic element. (b) Nodal forces and displacements of an elasto-plastic element with a rigid body at each end.

usually given by a nonlinear and asymmetric surface with regard to the internal forces, as illustrated in Fig. 2. However, in order to facilitate the treatment of the yield condition, the yield surface can be approximated by a linearized function, resulting in an underestimation of the strength of the member, as shown by the thick lines in Fig. 2. The plasticity condition of a cross-section can then be expressed in the following form:

$$F_k = R_k - \mathbf{C}_k^T \mathbf{X}_t = 0 \quad (k = i, j) \tag{1}$$

In Eqn. (1), R_k is the reference strength of the element end k, which is taken to be a fully plastic moment; i.e. $R_k = \sigma_{yk} AZ_{pk}$, where AZ_{pk} is the plastic section modulus of element end k, and σ_{yk} is the yield stress. $\mathbf{C}_k^T$ is a factor determined by the dimensions of element k. Particularly, an expression taking account of the bending moment, and the shearing and axial force effect upon the plasticity condition is given

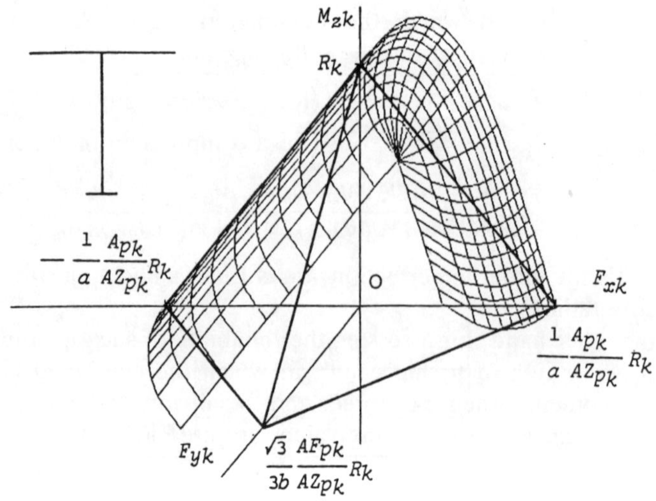

FIG. 2. Linearized plasticity conditions under the interaction of bending moment, shearing force and axial force.

as follows:

$$\mathbf{C}_i^T = (a \cdot AZ_{pi}/A_{pi}\ \mathrm{sign}(F_{xi}),\ b \cdot \sqrt{3}\,AZ_{pi}/AF_{pi}\ \mathrm{sign}(F_{yi}),\ \mathrm{sign}(M_{zi}),\ 0,\ 0,\ 0) \quad (1a)$$

$$\mathbf{C}_j^T = (0,\ 0,\ 0,\ a \cdot AZ_{pj}/A_{pj}\ \mathrm{sign}(F_{xj}),\ b \cdot \sqrt{3}\,AZ_{pj}/AF_{pj}\ \mathrm{sign}(F_{yj}),\ \mathrm{sign}(M_{zj})) \quad (1b)$$

where A_{pi}, A_{pj} is the cross-sectional area of the element end, AF_{pi}, AF_{pj} is the effective sectional area of the element end for the shearing force, sign$(\cdot)$ is the sign of $(\cdot)$ and a, b are the coefficients of the axial force and shearing force effects, respectively.

The plasticity condition in Eqn. (1) reduces to one of the following forms:

(1) In the case of $a = b = 0$, the condition is subject solely to the bending moment.
(2) In the case of $a \neq 0$ and $b = 0$, the condition is subject to the interaction of the bending moment and axial force.
(3) In the case of $a \neq 0$ and $b \neq 0$, the condition is subject to the interaction of the bending moment, shearing force and axial force.

For a structure in which the bending moment and axial force dominate the failure criterion, the following values are adopted [15]:

$$a = 1 \qquad b = 0 \qquad \text{(for a tension member)} \tag{2a}$$

$$a = \max[\sigma_{yk}/\sigma_{ck}, 1] \qquad b = 0 \qquad \text{(for a compression member)} \tag{2b}$$

$$\sigma_{ck} = (\sigma_{yk} + \sigma_E + (w_0/s)\sigma_E)/2 \times \{1 - \sqrt{1 - 4\sigma_E\sigma_{yk}/(\sigma_{yk} + \sigma_E + (w_0/s)\sigma_E)^2}\}$$

where w_0 is the initial imperfection, σ_E is Euler's buckling stress and s is the core radius.

On the other hand, for assessing the reliability of such a structure as the transverse ring of a ship's hull, in which the combined effect of bending moment, shearing force and axial force determines the plasticity condition, the following values are used [17]:

$$a = 1 \qquad b = 0{\cdot}5 \tag{3}$$

The behavior of the yielded section follows plastic theory because a perfectly elasto-plastic relationship has been employed in the plasticity condition. The relationship between the nodal force vector ($\mathbf{X}_t$) and the displacement vector ($\boldsymbol{\delta}_t$) of an element including plastic nodes can next be derived by using plastic theory as follows [9, 18];

$$\mathbf{X}_t = \mathbf{k}_t^{(p)}\boldsymbol{\delta}_t + \bar{\mathbf{X}}_t^{(p)} \tag{4}$$

where $\mathbf{k}_t^{(p)}$ is the reduced element stiffness matrix and $\bar{\mathbf{X}}_t^{(p)}$ is the equivalent nodal force vector. The explicit forms of $\mathbf{k}_t^{(p)}$ and $\bar{\mathbf{X}}_t^{(p)}$ can now be expressed.

(a) In the case of an elastic element:

$$\left.\begin{aligned} &\mathbf{k}_t^{(p)} = \mathbf{k}_t \ (\mathbf{k}_t \text{ is the elastic element stiffness matrix}) \\ &\bar{\mathbf{X}}_t^{(p)} = 0 \end{aligned}\right. \tag{5a}$$

(b) In the case of failure at the left-hand end:

$$\left.\begin{aligned} \mathbf{k}_t^{(p)} &= \mathbf{k}_t - \mathbf{k}_t\mathbf{C}_i\mathbf{C}_i^T\mathbf{k}_t/(\mathbf{C}_i^T\mathbf{k}_t\mathbf{C}_i) \\ \bar{\mathbf{X}}_t^{(p)} &= R_i\mathbf{k}_t\mathbf{C}_i/(\mathbf{C}_i^T\mathbf{k}_t\mathbf{C}_i) \end{aligned}\right. \tag{5b}$$

(c) In the case of failure at the right-hand end:

$$\left.\begin{aligned} \mathbf{k}_t^{(p)} &= \mathbf{k}_t - \mathbf{k}_t\mathbf{C}_j\mathbf{C}_j^T\mathbf{k}_t/(\mathbf{C}_j^T\mathbf{k}_t\mathbf{C}_j) \\ \bar{\mathbf{X}}_t^{(p)} &= R_j\mathbf{k}_t\mathbf{C}_j/(\mathbf{C}_j^T\mathbf{k}_t\mathbf{C}_j) \end{aligned}\right. \tag{5c}$$

(d) In the case of failure at both ends:

$$\mathbf{k}_t^{(p)} = \mathbf{k}_t - [H]^T[G^{-1}][H]$$

$$\bar{\mathbf{X}}_t^{(p)} = [H]^T[G^{-1}]\begin{Bmatrix} R_i \\ R_j \end{Bmatrix} \tag{5d}$$

$$[G^{-1}] = \begin{bmatrix} \mathbf{C}_i^T\mathbf{k}_t\mathbf{C}_i, & \mathbf{C}_i^T\mathbf{k}_t\mathbf{C}_j \\ \mathbf{C}_j^T\mathbf{k}_t\mathbf{C}_i, & \mathbf{C}_j^T\mathbf{k}_t\mathbf{C}_j \end{bmatrix}^{-1} \quad [H] = \begin{bmatrix} \mathbf{C}_i^T\mathbf{k}_t \\ \mathbf{C}_j^T\mathbf{k}_t \end{bmatrix}$$

Consider an element with rigid bodies at both ends, which is used as the idealized structure of the transverse ring in a tanker hull.

Let $(\mathbf{X}_t)_r$ and $(\boldsymbol{\delta}_t)_r$, respectively, denote the nodal force and displacement vectors of the outside of the unit element i, j with rigid bodies whose lengths are s_1 and s_2, as shown in Fig. 1(b). By using transformation matrix for $\boldsymbol{\tau}_t$ in $\boldsymbol{\delta}_t = \boldsymbol{\tau}_t(\boldsymbol{\delta}_t)_r$ of

$$\boldsymbol{\tau}_t = \begin{bmatrix} 1 & 0 & 0 & 0 & 0 & 0 \\ 0 & 1 & s_1 & 0 & 0 & 0 \\ 0 & 0 & 1 & 0 & 0 & 0 \\ 0 & 0 & 0 & 1 & 0 & 0 \\ 0 & 0 & 0 & 0 & 1 & -s_2 \\ 0 & 0 & 0 & 0 & 0 & 1 \end{bmatrix} \tag{6}$$

and the relationship between $\mathbf{X}_t$ and $\boldsymbol{\delta}_t$ for an elasto-plastic element, the following relationship is obtained:

$$(\mathbf{X}_t)_r = (\mathbf{k}_t^{(p)})_r(\boldsymbol{\delta}_t)_r + (\bar{\mathbf{X}}_t^{(p)})_r \tag{7}$$

where

$$(\mathbf{k}_t^{(p)})_r = \boldsymbol{\tau}_t^T\mathbf{k}_t^{(p)}\boldsymbol{\tau}_t$$

and

$$(\bar{\mathbf{X}}_t^{(p)})_r = \boldsymbol{\tau}_t^T\bar{\mathbf{X}}_t^{(p)}$$

Generation of Safety Margins and the Structural Failure Criterion

Consider a plane frame structure with n elements and a maximum of $3l$ loads applied to its l nodes. The failure criterion for the ith elasto-plastic element end is given by

$$Z_i = R_i - \mathbf{C}_i^T\mathbf{X}_t \leq 0 \tag{8}$$

The failure of a frame structure is defined as the occurrence of a large nodal displacement due to plastic collapse. A criterion for structural failure can then be given in the following manner: When any one element end yields, the internal forces are redistributed to the element ends. Similarly when some members, e.g. $r_1, r_2, \ldots, r_{p-1}$, have failed, stress analysis can be performed again and the stiffness equation of the element replaced by the corresponding reduced equation, e.g. Eqn. (4) or Eqn. (7). The reduced element stiffness matrices are then evaluated for all the failed elements, and they are assembled to obtain the total structural stiffness matrix:

$$(\mathbf{K}^{(p)})(\mathbf{d}) = (\mathbf{L}) + (\mathbf{R}^{(p)}) \tag{9}$$

where $(\mathbf{d})$ is the total nodal displacement vector referred to the global coordinate system, $(\mathbf{K}^{(p)}) = \sum_{k=1}^{n} \mathbf{T}_k^T \boldsymbol{\tau}_k^T \mathbf{k}_k^{(p)} \boldsymbol{\tau}_k \mathbf{T}_k$ is the reduced total structural stiffness matrix, $\mathbf{T}_k$ is the transformation matrix, $(\mathbf{L})$ is the vector of external loads and $(\mathbf{R}^{(p)}) = -\sum_{k=1}^{n} \mathbf{T}_k^T \boldsymbol{\tau}_k^T \bar{\mathbf{X}}_k^{(p)}$ is the equivalent nodal force vector referred to the global coordinate system. Finally, the nodal force vector $\mathbf{X}_t$ of the tth element is given by

$$\mathbf{X}_t = \boldsymbol{b}_t^{(p)}[(\mathbf{L}) + (\mathbf{R}^{(p)})] + \bar{\mathbf{X}}_t^{(p)} \tag{10}$$

where $\mathbf{b}_t^{(p)} = \boldsymbol{\tau}_t \mathbf{T}_t [(\mathbf{K}_t^{(p)})]^{-1}$ and $[(\mathbf{K}_t^{(p)})]^{-1}$ is the matrix formed by extracting the rows corresponding to the tth element from the matrix $[(\mathbf{K}^{(p)})]^{-1}$.

Now that the element ends $r_1, r_2, \ldots, r_{p-1}$ have failed, the safety margin of the surviving element end i (element number t) can be obtained by substituting Eqn. (10) into Eqn. (7):

$$Z_i^{(p)} = R_i + \mathbf{C}_i^T\left(\mathbf{b}_t^{(p)} \sum_{k=1}^{n} \mathbf{T}_k^T \boldsymbol{\tau}_k^T \bar{\mathbf{X}}_k^{(p)} - \bar{\mathbf{X}}_t^{(p)}\right) - \mathbf{C}_i^{(T)} \mathbf{b}_t^{(p)}(\mathbf{L}) \tag{11}$$

$$= R_i + \sum_{k=1}^{p-1} a_{ir_k}^{(p)} R_{r_k} - \sum_{j=1}^{3l} b_{ij}^{(p)} L_j \tag{12}$$

where $a_{ir_k}^{(p)}$ and $b_{ij}^{(p)}$ are the coefficients resulting from resolution of the vectors into their components.

The occurrence of large nodal displacements due to plastic collapse can be determined by investigating the property of the total structural stiffness matrix $(\mathbf{K}^{(p)})$. For example, when the element ends at a specific number p_q, i.e. element ends $r_1, r_2, \ldots, r_{p_q}$, have failed and the reduced total structural stiffness matrix $(\mathbf{K}^{(p_q)})$ satisfies the following condition, structural failure results:

$$|(\mathbf{K}^{(p_q)})|/|(\mathbf{K}^{(0)})| \leq \varepsilon \tag{13}$$

where superscripts (p_q) and (0) are used to denote the p_qth failure stage and the elastic condition, respectively. ε is the specified constant for determining the plastic collapse.

By using Eqn. (13), the criterion for structural failure can be given by

$$Z_{r_p}^{(p)} \leq 0 \quad (p = 1, 2, \ldots, p_q) \tag{14}$$

SELECTION OF PROBABILISTICALLY DOMINANT FAILURE PATHS

There are too many failure paths in a highly redundant structure [18, 19] to generate all of them, which necessitates a procedure for selecting only the probabilistically significant failure paths. Efficient methods using a branch-and-bound technique have been proposed, [5–8, 15, 16, 18, 19] and this paper adoptes this technique according to the subsequent procedure.

Branching Operation

These operations are used to select the plastic nodes so that stochastically dominant failure paths may be obtained. An element end (called here a section for simplicity) is selected as a plastic node at the pth failure stage based on two-dimensional joint probability (so-called two-dimensional branchings). The section to be selected at the pth failure stage is given by

$$P[Z_{r_1}^{(1)} \leq 0] = \max_{i_1 \in I_1} P[Z_{i_1}^{(1)} 0] \qquad \text{for } p = 1 \tag{15}$$

$$P[(Z_{r_1}^{(1)} \leq 0) \cap (Z_{r_p}^{(p)} \leq 0)] = \max_{i_p \in I_p} P[(Z_{r_1}^{(i)} \leq 0) \cap (Z_{i_p}^{(p)} \leq 0)] \qquad \text{for } p \geq 2 \tag{16}$$

where I_p is the set of sections i_p to be selected at the pth failure stage, $Z_{i_1}^{(1)}$ is the safety margin of section i_1 at the first failure stage; i.e. when no plastic nodes exist in the structure and $Z_{i_p}^{(p)}$ is the safety margin of section i_p at the pth failure stage, i.e. after the formation of plastic nodes at sections $r_1, r_2, \ldots, r_{p-1}$ $(p \geq 2)$.

The joint probability can be calculated by the Hermite polynomial expansion method [4]. By repeating the selection process, a sequence of plastic node sections to form plastic collapse, i.e. a complete failure path $(r_1, r_2, \ldots, r_{p_q})$, can be found.

The lower and upper bounds, $P^{(p)}_{fp(q)(L)}$ and $P^{(p)}_{fp(q)(U)}$, of the probability $P^{(p)}_{fp(q)}$ of a partial failure path up to the pth ($p \geq 2$) failure stage can be evaluated by the following equations [19]:

$$P^{(p)}_{f(q)(L)} \leq P^{(p)}_{fp(q)} = P\left[\bigcap_{i=1}^{p} (Z^{(i)}_{r_i(q)} \leq 0)\right] \leq P^{(p)}_{fp(q)(U)} \tag{17}$$

$$P^{(p)}_{fp(q)(U)} = \min_{j \in \{2, \ldots, p\}} P[(Z^{(1)}_{r_1(q)} \leq 0) \cap (Z^{(j)}_{r_j(q)} \leq 0)] \tag{18}$$

$$P^{(p)}_{fp(q)(L)} = \max\left\{0,\ 1 - P[S_1] - \sum_{i=2}^{p} \min_{j \in \{1, 2, \ldots, i-1\}} P[\bar{S}_j \cap S_i]\right\} \tag{19}$$

In Eqn. (19), S_i terms designate the nonfailure events $Z^{(i)}_{r_i(q)} > 0$ ($i = 1, 2, \ldots, p$) rearranged in the decreasing order of probability [4]:

$$P[S_1] \geq P[S_2] \geq \cdots \geq P[S_p] \tag{20}$$

Further, the following bound [20] is also applicable when all the correlation coefficients are nonnegative, i.e.

$$\begin{gathered} P^{(p)}_{fp(q)(L)} = \int_{-\infty}^{\infty} \phi(t) \cdot \prod_{j=1}^{p} \Phi((-\beta_j - t\lambda_j)/(1 - \lambda_j^2)^{1/2})\, dt \\ \lambda_j = \left[\min_{i \neq j} \{\rho_{ij}\}\right]^{1/2}, \qquad \{\rho_{ij}\} \geq 0 \end{gathered} \tag{21}$$

β_j is the reliability index at the jth failure stage, and ϕ and Φ are the standard normal probability density function and standard normal probability distribution function, respectively.

Equations (19) and (21) need a safety margin at each failure stage. It should be noted here that the lower bound is calculated only when a complete failure path has been found.

The maximum P_{fpM} of the lower bounds of the selected complete failure path probability is calculated from

$$P_{fpM} \triangleq \max_{q} P^{(p_q)}_{fp(q)(L)} \tag{22}$$

P_{fpM} is updated when a new complete failure path has been found with a failure probability larger than the previous P_{fpM}. The branching operations are terminated when no sections are left for selection.

Bounding Operations

These operations are used to select the sections to be discarded. The sections deleted at the pth failure stage are

$$P[Z_{i_1}^{(1)} \leq 0]/P_{fpM} < 10^{-\gamma_1} \quad \text{for} \quad p = 1 \tag{23}$$

$$P[(Z_{r_1}^{(1)} \leq 0) \cap (Z_{i_p}^{(p)} \leq 0)]/P_{fpM} < 10^{-\gamma_2} \quad \text{for} \quad p \geq 2 \tag{24}$$

where γ_1 and γ_2 are the specified constants.

From this, it can be concluded that the failure paths to be neglected are those which have a failure probability smaller than $10^{-\gamma_i} P_{fpM}$ ($i = 1, 2$).

The probability of occurrence P_{fp} for the failure mode, i.e. the set of plastic nodes to produce structural failure, corresponding to the selected failure path, can be estimated by

$$P_{fq} = P[Z_{r_{pq}}^{(pq)} \leq 0] \tag{25}$$

Heuristic Operations

The number of branchings becomes enormous for a large-scale structure with a high degree of redundancy, even though the branch-and-bound method is applied. To reduce the computational effort, three heuristic operations can be applied. First, a reliability assessment is performed on some structural divisions that are presumed to be critical. Second, the set of sections for branching is restricted to the sections which satisfy the monotonic conditions of failure probability. That is

$$I_p = \{i_p \mid P[(Z_{r_1}^{(1)} \leq 0) \cap (Z_{i_p}^{(p)} \leq 0)] \leq P[(Z_{r_1}^{(1)} \leq 0) \cap (Z_{r_{p-1}}^{(p-1)} \leq 0)]\} \tag{26}$$

Third, the number of branchings from one failure stage is restricted to a specified number.

NUMERICAL EXAMPLES

The proposed method is now applied to a transmission line tower [16] and to the transverse ring of a large tanker hull [17].

Transmission Line Tower

Consider the structure of a transmission line tower shown in Fig. 3. The dimensions and strengths of members are listed in Table 1, and the strengths of the nodes are assumed to be mutually independent

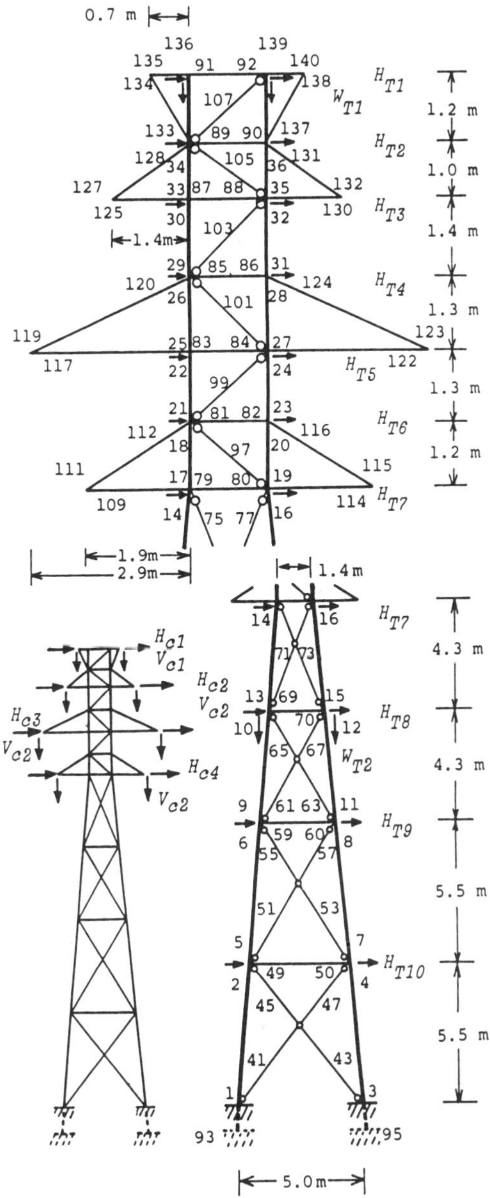

FIG. 3. Structure of a transmission line tower.

TABLE 1
Numerical data for the transmision line tower

Element end number	*Outside diameter D_i (mm)*	*Thickness t_i (mm)*
1–8	190·7	16·0
9–12	190·7	14·0
13–16	165·2	12·0
17–24	165·2	8·0
25–28	139·8	5·5
29–32	139·8	4·5
33–36	114·3	3·2
37–40	101·6	3·2
41–48 51–58 61–68	89·1	5·0
71–78	101·6	6·0
49, 50, 59 60, 69, 70	65·5	2·6
79–82 87–90	114·3	2·9
83–86	114·3	3·2
91–92	48·6	2·4
97–100	139·8	5·5
101–104	114·3	4·0
105–106 109–116	89·1	3·2
107–108 125–132	89·1	2·9
117–124	114·3	3·5
133–140	89·1	2·6

Young's modulus $E = 210$ GPa
Mean value of yield stress
$\bar{\sigma}_{yi} = 400$ MPa $(i = 1, 2, \ldots, 40)$
$= 276$ MPa $(i = 41, 42, \ldots, 140)$
Coeff. of variation $CV_{\sigma_{Yi}} = 0{\cdot}12$
Mean value of foundation pile (93–96) capacity
$\bar{R}_f = 22.4$ MN
Coeff. of variation $CV_{R_f} = 0{\cdot}15$
Correlation coeff. $\rho = 0{\cdot}5$ for all the element ends
All brace elements are truss

TABLE 2
Applied loads on the transmission line tower (see Fig. 3)

	H_{C1}	H_{C2}	H_{C3}	H_{C4}	V_{C1}	V_{C2}	W_{T1}	W_{T2}
Mean value (kN)	17·30	24·73	24·51	24·29	13·01	18·14	3·59	14·89
Coefficient of variation	0·39	0·37	0·37	0·37	0·41	0·38	0·02	0·02

	H_{T1}	H_{T2}	H_{T3}	H_{T4}	H_{T5}	H_{T6}	H_{T7}	H_{T8}	H_{T9}	H_{T10}
Mean value (kN)	0·13	0·25	0·30	0·34	0·35	0·36	0·68	1·04	1·24	2·05

Coefficient of variation $CV_{H_{Ti}}$ $(i = 1, 2, \ldots, 10) = 0{\cdot}37$
Correlation coefficient $\rho_{H_{Ci}H_{Cj}} = \rho_{H_{Ci}V_{Cj}} = \rho_{H_{Ti}H_{Tj}} = 1{\cdot}0$, $\rho_{mn} = 0{\cdot}0$ $(m, n$: others).

normal random variables. The data for external loadings are also given in Fig. 3 and Table 2. The plasticity condition takes account of the combined loading effect of bending moment and axial force ($a = \sigma_{yk}/\sigma_{ck}$, $b = 0$). The tower is divided into eight structural divisions and foundation, which are presumed to be critical. The number of branchings from each failure stage has been limited to 2, the results being listed in Table 3.

It can be seen that the dominant failure mode is the side-sway mechanism of the fifth division. This example demonstrates that selecting the dominant failure mode and assessing the reliability of the structure of a transmission line tower can be performed in a very short computation time by using heuristic operations.

Transverse Ring of a Large Tanker Hull

Figure 4 shows the plane frame structure that has been modelled for the transverse ring of a large tanker hull under two notional loading conditions. The probabilistic analysis of the plastic collapse was carried out, using the numerical data for the structure given in Fig. 4 and Table 4. The applied loads were estimated, based on the loading condition for direct calculation suggested by the Japan Classification Society for Ships (NK), and the lengths of rigid bodies were estimated by using 'the span point for bending' given in Ref. 21. It was assumed

TABLE 3

Selected failure modes and their probabilities of occurrence for the transmission line tower

$\varepsilon = 0{\cdot}001,\ \gamma_1 = \gamma_2 = 0{\cdot}0$

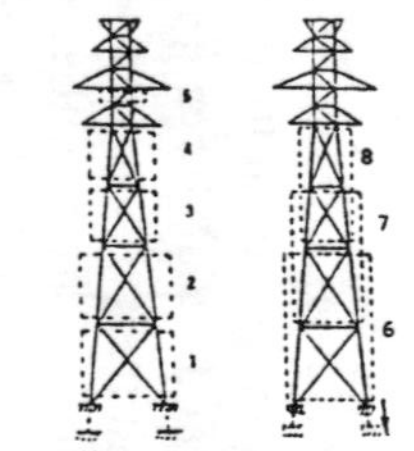

Division	*Failure modes*	*Failure probability P_{f_q}*	*Division*	*Failure modes*	*Failure probability P_{f_q}*
Fifth	(21, 22, 23, 24, 99)[a]	0·2616 $\times 10^{-7}$ [2][b]	Eighth	(13, 14, 15, 16, 71, 73)	0·1740 $\times 10^{-9}$ [2]
Computation time		145·6 s		(10, 12, 14, 16, 69, 70, 71, 73)	—
Fourth	(13, 14, 15, 16, 71, 73)	0·1754 $\times 10^{-9}$ [2]		(9, 11, 13, 15, 61, 63, 69, 70)	—
	(13, 14, 15, 16, 75, 77)	—	Computation time		1391·4 s
Computation time		272.9 s	Seventh	(6, 8, 10, 12, 59, 60, 65, 67)	0·2412 $\times 10^{-16}$ [1]
Third	(9, 10, 11, 12, 61, 63)	—		(8, 9, 10, 12, 60, 61, 65)	0·1661 $\times 10^{-16}$ [1]
Computation time		1157.7 s		(8, 11, 60)	0·1343 $\times 10^{-16}$ [2]
Second	(5, 6, 7, 8, 51, 53)	0·2026 $\times 10^{-19}$ [1]		(6, 8, 10, 12, 59, 60, 61, 63)	—
	(5, 6, 7, 8, 55, 57)	—		(9, 10, 11, 12, 61, 63)	0·4291 $\times 10^{-17}$ [1]
Computation time		815·0 s		(5, 6, 7, 8, 51, 53)	—
First	(1, 2, 3, 4, 41, 43)	0.3476 $\times 10^{-20}$ [3]	Computation time		2036·2 s
	(1, 2, 3, 4, 43, 47)	—	Sixth	(1, 3, 6, 8, 43, 47, 49, 50, 51, 53)	0·1177 $\times 10^{-18}$ [1]
	(1, 2, 3, 4, 41, 45)	—		(2, 6, 7, 8, 49, 51, 53)	0·2805 $\times 10^{-19}$ [3]
Computation time		531·6 s		(1, 3, 6, 8, 45, 47, 49, 50, 51, 53)	—
				(1, 3, 6, 8, 43, 47, 49, 50, 53, 57)	—
			Others		$<$0·6 $\times 10^{-22}$ [1]
			Computation time		2544·6 s
			Foundation (95: overturning	0·1105 $\times 10^{-9}$	
			Computation time		29·1 s

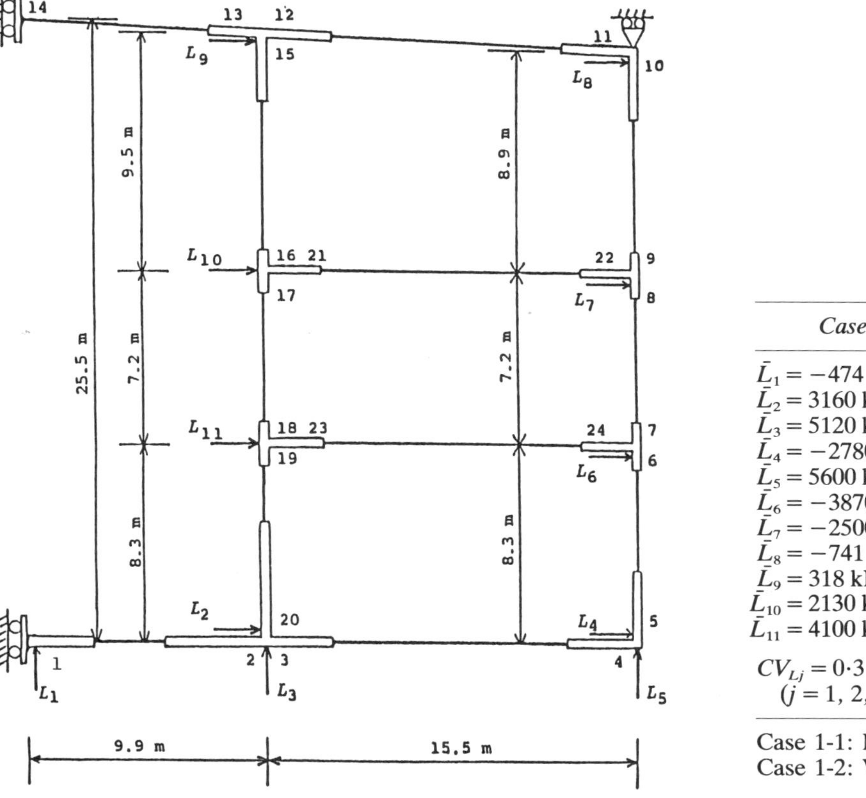

Case 1-1	Case 1-2
$\bar{L}_1 = -474$ kN	$\bar{L}_1 = 2920$ kN
$\bar{L}_2 = 3160$ kN	$\bar{L}_2 = -3160$ kN
$\bar{L}_3 = 5120$ kN	$\bar{L}_3 = 2020$ kN
$\bar{L}_4 = -2780$ kN	$\bar{L}_4 = 385$ kN
$\bar{L}_5 = 5600$ kN	$\bar{L}_5 = -905$ kN
$\bar{L}_6 = -3870$ kN	$\bar{L}_6 = 230$ kN
$\bar{L}_7 = -2500$ kN	$\bar{L}_7 = -370$ kN
$\bar{L}_8 = -741$ kN	$\bar{L}_8 = -420$ kN
$\bar{L}_9 = 318$ kN	$\bar{L}_9 = -318$ kN
$\bar{L}_{10} = 2130$ kN	$\bar{L}_{10} = -2130$ kN
$\bar{L}_{11} = 4100$ kN	$\bar{L}_{11} = -4100$ kN
$CV_{Lj} = 0{\cdot}3$ $(j = 1, 2, \ldots, 12)$	$CV_{Lj} = 0{\cdot}3$ $(j = 1, 2, \ldots, 12)$

Case 1-1: Main tank fully loaded.
Case 1-2: Wing tank fully loaded.

FIG. 4. Transverse ring of a large tanker hull (2400 MN DW) and the notional loading conditions.

TABLE 4
Numerical data for the transverse ring of a large tanker hull

Element and number	*Cross-sectional area* $A_i = A_{pi}$ (m^2)	*Cross-sectional area of web* $A_{wi} = AF_{pi}$ (m^2)	*Moment of inertia* I_i (m^4)	*Mean value of reference strength* $\bar{R}_i$ *(kNm)*	*Length of rigid body* s_{1i} *(m)*	s_{2i} *(m)*
1, 2	0·183	0·064	0·438	59 200·0	2·8	4·2
3, 4	0·176	0·054	0·255	43 100·0	2·8	2·9
5, 6	0·135	0·045	0·156	29 400·0	3·1	1·0
7, 8	0·135	0·045	0·156	29 400·0	1·0	1·0
9, 10	0·126	0·035	0·147	25 900·0	1·0	3·0
11, 12	0·136	0·030	0·076	14 600·0	3·0	2·9
13, 14	0·137	0·030	0·079	15 100·0	2·2	0·0
15, 16	0·089	0·035	0·117	24 800·0	3·0	0·9
17, 18	0·099	0·035	0·127	25 700·0	0·9	1·0
19, 20	0·116	0·045	0·142	29 200·0	1·0	5·1
21, 22	0·043	0·018	0·016	6650·0	2·3	2·3
23, 24	0·056	0·025	0·026	9700·0	2·3	2·3

Young's modulus $E = 210$ GPa.
Mean value of yield stress $\bar{\sigma}_{Yi} = 276$ MPa.

that the strengths of the nodes and the applied loads were mutually independent normal random variables.

The results for the Case 1-1 loading condition are listed in Table 5, the first column indicating the selected failure paths. In the second column, the probabilities for the occurrence of failure paths are given when the combined effect of the bending moment, shearing force and axial force ($a = 1,\ b = 0{\cdot}5$) was considered. The numbers in brackets are those of the selected failure paths. The third column shows the probabilities corresponding to the case in which the combined effect of the bending moment and axial force was considered ($a = 1$, $b = 0$), and the fourth column shows those corresponding to the case in which only the bending moment effect was considered ($a = b = 0$). The collapse modes are given in the fifth column. In each column, the number in parentheses indicates the central safety factor (CSF) corresponding to each failure path.

$$\mathrm{CSF} = \left(\bar{R}_i + \sum_{k=1}^{p_q-1} a_{ir_k}^{(p_q)} \bar{R}_{r_k}\right) \Big/ \sum_{j=1}^{3l} b_{ij}^{(p_q)} \bar{L}_j \qquad (27)$$

TABLE 5
Failure modes and their probabilities of occurrence for the transverse ring of a large tanker hull (Case 1-1 loading condition)
$\varepsilon = 0{\cdot}001,\ \gamma = 0,\ CV_{Rj}/CV_{Lj} = 0{\cdot}05/0{\cdot}3$

No.	Failure paths	Failure probability P_{fq}: *Bending moment, axial force and shearing force interaction considered* ($a = 1,\ b = 0{\cdot}5$)	Failure probability P_{fq}: *Bending moment and axial force interaction considered* ($a = 1,\ b = 0$)	Failure probability P_{fq}: *Bending moment only* ($a = 0,\ b = 0$)	Collapse mode
A	(5, 12, 22, 11, 24, 3, 23, 21,				
−1.		18) $0{\cdot}7482 \times 10^{-2}$ [60] (1·675)[a]	2, 18) $0{\cdot}3625 \times 10^{-5}$ [94] (2·265)	2, 18) 0.6683×10^{-8} [131] (2·945)	
−2.		8, 18) $0{\cdot}5523 \times 10^{-2}$ [60] (1·699)	1, 18) $0{\cdot}2574 \times 10^{-5}$ [35] (2·307)	1, 18) $0{\cdot}5453 \times 10^{-8}$ [22] (2.970)	
−3.		2) $0{\cdot}5481 \times 10^{-2}$ [12] (1·709)	2, 8, 18) $0{\cdot}1941 \times 10^{-5}$ [40] (2·273)	2, 19) $0{\cdot}7446 \times 10^{-9}$ [77] (3·071)	
−4.		others $<0{\cdot}5 \times 10^{-2}$ [95]	2, 8, 19) $0{\cdot}1693 \times 10^{-5}$ [33] (2·398)	1, 19) $0{\cdot}6701 \times 10^{-9}$ [22] (3·093)	
B	(5, 12, 22, 24, 10, 3, 23, 21, 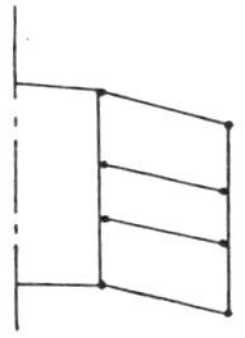				
−1.		18) $0{\cdot}6573 \times 10^{-2}$ [12] (1·658)	2, 18) $0{\cdot}4908 \times 10^{-5}$ [32] (2.130)	—	
−2.		8, 18) $0{\cdot}5903 \times 10^{-2}$ [2] (1·667)	1, 18) $0{\cdot}2414 \times 10^{-5}$ [16] (2.184)	—	
−3.		others $<0{\cdot}4 \times 10^{-2}$ [11]	others $<0{\cdot}1 \times 10^{-5}$ [9]	—	
C	(21, 12, 11, 24, 3, 4, 23, 22,				
−1.		—	5, 8) $0{\cdot}6574 \times 10^{-7}$ [1] (2·435)	—	
D	Others	$<0{\cdot}1 \times 10^{-3}$ [10]	$<0{\cdot}3 \times 10^{-7}$ [22]	$<0{\cdot}1 \times 10^{-9}$ [48]	
Total number of selected paths		[262]	[282]	[300]	
Computation time (s)		190·8	169·3	172·8	

[a] Central safety factor $= (R_i + \sum_{k=1}^{p_q-1} a_{ir_k}^{(p_q)} \bar{R}_{r_k}) / \sum_{j=1}^{3l} b_{ij}^{(p_q)} \bar{L}_j$.

TABLE 6

Most dominant collapse mode based on a probabilistic analysis under two notional loading conditions for the transverse ring of a large tanker hull $CV_{\sigma_{Yi}}/CV_{L_j} = 0{\cdot}05/0{\cdot}3$

Type of structure and notional loading condition	*Most dominant collapse mode and its probability of occurrence*		
	Coeff. of combined effect	*Failure probability*	*Most dominant collapse mode*
Tanker 2400 MN DW — *Case 1-1* 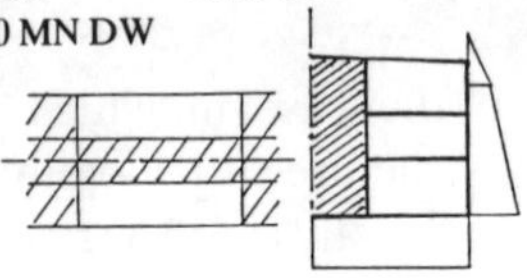	$a = 1, b = 0{\cdot}5$	$0{\cdot}7482 \times 10^{-2}$ (1·68)[a]	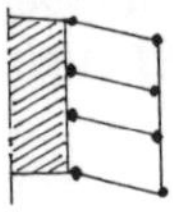
	$a = 1, b = 0$	$0{\cdot}3625 \times 10^{-5}$ (2·27)	
	$a = 0, b = 0$	$0{\cdot}7446 \times 10^{-8}$ (2·95)	
Case 1-2 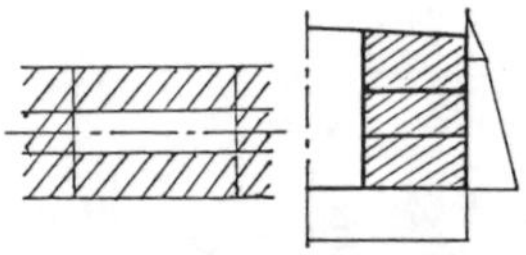	$a = 1, b = 0{\cdot}5$	$0{\cdot}3896 \times 10^{-4}$	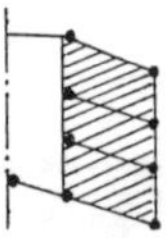
	$a = 1, b = 0$	$0{\cdot}1640 \times 10^{-10}$	
	$a = 0, b = 0$	$0{\cdot}5120 \times 10^{-13}$	

[a] Values in parentheses indicate the central safety factor in Eqn. (27).

It can be seen from Table 5 that the dominant failure mode for each case is essentially similar, being formed by failure of the wing tank. However, the probabilities of occurrence with a combined loading effect considered are very large, as can be seen by the failure paths A-1 to A-4 in the table. Moreover, it can be judged from a comparison between the central safety factor and failure probabilities that share the same failure path that the deterministically dominant path is not always stochastically relevant.

Finally, Table 6 shows the most dominant collapse mode based on a probabilistic analysis for the transverse structure of a large tanker hull under two notional loading conditions. It can be seen from this table that the full-load condition with an empty wing tank, Case 1-1, is the most severe.

CONCLUDING REMARKS

Methods have been presented for the reliability assessment of large-scale frame structures based on the plastic collapse analysis.

These methods were applied to the structure of a transmission line tower and of the transverse ring of a tanker hull under some notional loading conditions. In the former example, the applicability of heuristic procedures were demonstrated for selecting the probabilistically dominant failure modes and for assessing the reliability of large-scale frame structures. In the latter example, it was concluded that the linearized plasticity condition can be effectively applied to consider the combined loading effect on the plastic collapse analysis of a structure consisting of comparatively deep girders; in such a case, the failure probabilities with the shearing effect considered were significantly larger than those excluding this effect.

Although this analysis has been confined to the case in which a structural system was idealized as a plane frame structure, it is possible for this method to be extended to the reliability analysis of a space frame structure by incorporating terms for the biaxial bending moment, torsional moment, etc. in the plasticity conditions of Eqn. (1).

REFERENCES

[1] O. Ditlevsen and P. Bjerager, *Structural Safety,* **4**(3), 195 (1986).

[2] M. R. Gorman, Case Western Reserve University, Report No. 79-2 (1979).

[3] A. H. Ang and H. F. Ma, *Structural Safety and Reliability,* T. Moan and M. Shinozuka, Eds., Elsevier, Amsterdam, 295 (1981).

[4] Y. Murotsu, M. Yonezawa, F. Oba and K. Niwa, *Advances in Reliability and Stress Analysis,* J. J. Burns, Jr. Ed., ASME, New York, 3 (1979).

[5] Y. Murotsu, H. Okada, K. Niwa and S. Miwa, *Trans. ASME, J. Mech. Design,* **102**(4), 749 (1980).

[6] Y. Murotsu, H. Okada, K. Niwa and S. Miwa, *Reliability, Stress Analysis and Failure Prevention Methods in Mechanical Design,* M. D. Milestone, Ed., ASME, New York, 81 (1980).

[7] Y. Murotsu, H. Okada, M. Yonezawa and K. Taguchi, *Structural Safety and Reliability,* T. Moan and M. Shinozuka, Ed., Elsevier, Amsterdam, 315 (1981).

[8] Y. Murotsu, *Reliability Theory and Its Application in Structures and Soil Mechanics,* P. Thoft-Christensen, Ed., Martinus Nijhoff, The Hague, 525 (1982).

[9] H. Okada, S. Matsuzaki and Y. Murotsu, *Bull. Univ. Osaka Pref., Ser. A,* **32**(2), 155 (1983).

[10] Y. Murotsu, H. Okada, M. Grimmelt, M. Yonezawa and K. Taguchi, *Structural Safety,* **2**(2), 17 (1984).

[11] Y. Murotsu, H. Okada, M. Yonezawa and M. Kishi, *Fourth ICASP in Soil and Structural Engineering,* Universitat di Firenze, (Italy), Pitagora Editrice, Florence, 1325 (1983).

[12] R. E. Melchers and L. K. Tang, *Structural Safety,* **2**(2), 127 (1984).
[13] Y. Murotsu, M. Kishi, H. Okada, Y. Ikeda, and S. Matsuzaki, *Proc. 4th International OMAE Symposium,* J. S. Chung, V. J. Lunardini, S. K. Chakrabarti, Y. S. Wang, D. S. Sodhi and K. Karal, Eds., ASME, New York, **I,** 250 (1985).
[14] Y. Murotsu, H. Okada and S. Matsuzaki, *Structural Safety and Reliability,* I. Konishi, A. H. Ang and M. Shinozuka, Eds., IASSAR, Columbia University, NewYork, **I,** 117 (1985).
[15] Y. Murotsu, H. Okada, S. Matsuzaki and S. Katsura, *Proc. 5th OMAE,* ASME, New York, **II,** 9 (1986).
[16] Y. Murotsu, H. Okada, S. Matsuzaki and S. Nakamura, *Proceedings of the International Symposium on Probability Methods Applied to Electric Power Systems,* S. G. Krishnasamy, Ed., Pergamon Press, Toronto, 53 (1986).
[17] H. Okada, Y. Murotsu, S. Matsuzaki and S. Katsura, *J. Soc. Naval Architects Japan,* **159,** 239 (1986), (in Japanese).
[18] P. Thoft-Christensen and Y. Murotsu, *Application of Structural Systems Reliability Theory,* Springer-Verlag, Berlin, (1986).
[19] Y. Murotsu, S. Matsuzaki and H. Okada, *JSME Int. J.,* **30**(260), 234 (1987).
[20] M. Hohenbichler and R. Rackwitz, *Structural Safety,* **1**(3), 177 (1983).
[21] I. Yamaguchi, *J. Soc. Naval Architects Japan,* **109,** 213 (1961) (in Japanese).

Reliability Analysis of Damaged Redundant Structures

Hitoshi Furuta[a], Masaaki Ohshima[b]* and Naruhito Shiraishi[a]

[a] *Department of Civil Engineering, Kyoto University, Kyoto, Japan*

[b] *Kounoike-gumi Corporation, Osaka, Japan*

ABSTRACT

The concept of fuzzy sets is applied to the reliability analysis of damaged redundant structures. The damaged state is represented in terms of fuzzy sets which specify such conditions as 'severely damaged', 'moderately damaged' and 'slightly damaged'. In formulating a reliability analysis, information regarding the damaged state can be introduced by coupling with the PNET method. While the PNET method is useful for calculating the system reliability of structures, it is necessary to identify all the possible failure mechanisms, a task which becomes more difficult as the number of failure mechanisms increases. To overcome this problem, a technique proposed by Ishikawa *et al.* is used, which can provide sufficient dominant failure mechanisms. A numerical examples is presented to illustrate the method.

INTRODUCTION

In order to minimize the risk of catastrophic failures causing social and economic losses, daily maintenance is important and inevitable. To establish an appropriate maintenance and repair program, it is necessary to evaluate the reliability of existing structures, most of which suffer from damage resulting from corrosion, cracking and other types of deterioration [1–3]. However, analyzing the reliability of a damaged structure is not an easy task due to the lack of statistical data available and the difficulty in estimating the damaged state of the structure under consideration [4–6]. So far, the reliability analyses of existing structures have been performed based on the judgment and intuition of experienced engineers. Although this method is simple and

* Formerly at Kyoto University.

practical, it has neither a theoretical basis nor consistency between structures.

In this paper, an attempt is made to derive a simple but well-grounded method for evaluating the reliability of damaged structures. Especially, attention is paid to the reliability assessment of redundant structures, because the relationship between the damaged states and the system reliability can be considered interesting and important from the standpoint of practical design.

The damaged state is defined in terms of linguistic variables which are specified by fuzzy sets [7]. By using a reduction factor, information regarding the damaged state can be introduced into the calculation of failure probability. Since the reduction factor can also be expressed by a fuzzy set, the failure probability obtained becomes a fuzzy set. This provides a useful insight for evaluating the resistance of each member to damage and the effect of damage level on the change of dominant failure modes. The system failure probability in this analysis is calculated by using the PNET method [8], which enables the system failure probability to be obtained without a complex calculating procedure. However, as the degree of structural redundancy increases, it becomes difficult to identify all the possible failure mechanisms. To overcome this difficulty, a technique proposed by Ishikawa *et al.* [9] is used, which can provide sufficient dominant failure mechanisms to estimate the system failure probability with accuracy. A numerical example will be presented to illustrate the applicability of the method developed here.

RELIABILITY ANALYSIS OF DAMAGED STRUCTURES

According to Yao [10], damage measurement can be generally classified into three types; numerical measurement, monetary measurement, and verbal measurement, the most desirable apparently being the numerical measurement. However, it is difficult, because of technical and financial constraints concerning the particular properties of civil engineering structures, to establish a useful numerical measurement for practical use.

The information sources available for damage assessment include design documents or drawings, visual inspections, field testing, laboratory testing, and structural analysis [11]. However, not all the available information can generally be used due to cost constraints. Actual daily

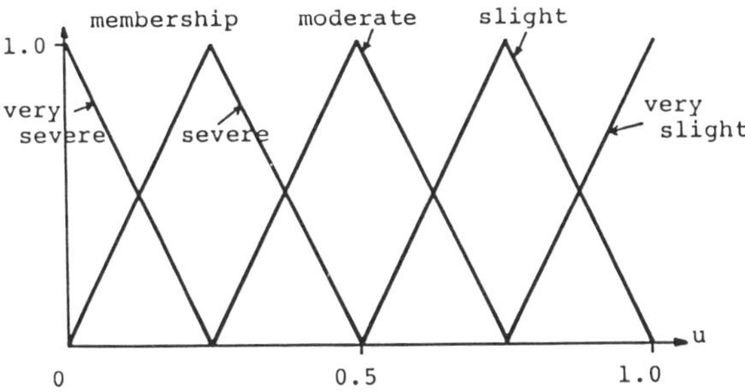

FIG. 1. Membership functions for verbal evaluations.

maintenance work is carried out on the basis of the intuition and engineering judgment of experienced inspectors. This judgment may possibly be derived from subjective feelings and insufficient vague data obtained from limited inquiries.

An attempt is made in this paper to apply the concept of fuzzy sets so as to introduce the subjective assessment of engineers into the reliability analysis of damaged structures. The damaged state of a structure is evaluated by a verbal measurement [12] such as 'very severe', 'severe', 'moderate', 'slight', and 'very slight'. These verbal evaluations are represented by reduction factors that can be defined by fuzzy sets as shown in Fig. 1. The abscissa (u) ranges between $u = 1$, denoting no damage, and $u = 0$, denoting failure. For simplicity, the shape of the membership functions are assumed to be triangular. When the damaged state can be estimated with accuracy, the function becomes sharp, but is otherwise flat.

Suppose that damage does not affect the load (S) but does affect the member resistance (R). Then, the safety margin Z can be written as

$$Z = \tilde{\phi}R - S \tag{1}$$

where $\tilde{\phi}$ is a reduction factor defined by a fuzzy set [13, 14], the symbol $\sim$ denoting a fuzzy quantity. It is noted that the safety margin Z in Eqn. (1) is a fuzzy quantity because $\tilde{\phi}$ is a fuzzy quantity. Assuming that R and S are statistically independent and normally distributed, the safety index $\tilde{\beta}$ can be calculated as

$$\tilde{\beta} = \frac{\tilde{\phi}\mu_R - \mu_S}{\sqrt{\tilde{\phi}^2\sigma_R^2 + \sigma_S^2}} \tag{2}$$

Then, the failure probability P_f is obtained as follows:

$$\tilde{P}_f = \Phi(-\tilde{\beta}) \tag{3}$$

where $\Phi(\ )$ is the normal distribution function, and μ_R, μ_S, σ_R and σ_S are the mean values and standard deviations of R and S, respectively. It should be noted that both Φ and P_f in Eqns. (2) and (3) are obtained as fuzzy quantities, so that normal arithmetic operations can be no longer used for these calculations. Here, a calculating method called the 'extension principle' [15] is employed for implementing Eqns. (2) and (3).

In order to obtain a new quantity C from A and B through the operation

$$\tilde{C} = f(\tilde{A}, \tilde{B}) \tag{4}$$

the membership function of $\tilde{C}$ is calculated as follows, based on the extension principle:

$$\mu_C(z) = \sup_{z=f(x,\,y)} \min(\mu_A(x), \mu_B(y)). \tag{5}$$

Although this principle is conceptually applicable to all cases including continuous and discrete variables, its implementation is not easy, even if a digital computer is used. Therefore, we will utilize an approximate calculation for the L and R functions that was developed by Dubois and Prade [16].

The representative arithmetic operations are shown as follows:

$$\oplus \quad (m, \alpha, \beta) \oplus (n, \gamma, \delta) = (m + n, \alpha + \gamma, \beta + \delta) \tag{6}$$

$$\ominus \quad (m, \alpha, \beta) \ominus (n, \gamma, \delta) = (m - n, \alpha + \delta, \beta + \gamma) \tag{7}$$

$$\odot \quad (m, \alpha, \beta) \odot (n, \gamma, \delta) = (mn, m\gamma + n\alpha, m\delta + n\beta) \tag{8}$$

$$\oslash \quad (m, \alpha, \beta) \oslash (n, \gamma, \delta) = (m/n, (m\delta + n\alpha)/n^2, (m\gamma + n\beta)/n^2) \tag{9}$$

The symbols $\oplus$, $\ominus$, $\odot$, $\oslash$ correspond to the usual operations of $+$, $-$, $\cdot$ and $\div$, respectively. In these expressions, a fuzzy quantity is expressed by three parameters; for instance, m, α and β denote the central value, and the scatters on left-hand side and right-hand side, respectively. Using the expression for the L and R functions, any membership function can be defined as follows:

$$\mu(x) = \begin{cases} L((m - x)/\alpha) & \text{for} \quad x \leq m,\ \alpha > 0 \quad (10) \\ R((x - m)/\beta) & \text{for} \quad x \geq m,\ \beta > 0 \quad (11) \end{cases}$$

In order to calculate Eqn. (2), it is necessary to define a new operation for the square root. Here, we derive a formula for this operation on the basis of the L and R functions [16].

$$\textcircled{\scriptsize V} \quad \sqrt{(m, \alpha, \beta)} \cong (\sqrt{m}, \alpha/2\sqrt{m}, \beta/2\sqrt{m}) \tag{12}$$

It is to be noted that this formula can be applied for cases with the same-shaped membership functions.

CALCULATION OF THE SYSTEM FAILURE PROBABILITY

In order to calculate the system reliability of redundant structures, it is necessary to consider the correlation between every failure mechanism. In this analysis, the PNET method proposed by Ang and Ma [8] is employed to obtain the system failure probability. While the PNET method is useful for calculating the system reliability of any type of structure, it requires all the possible failure mechanisms to be identified. As the number of failure mechanism increases, it becomes more difficult to achieve the possible mechanisms [17]. To overcome this problem, we will employ a technique proposed by Ishikawa and co-workers [9] which can provide a sufficient number of dominant failure mechanisms.

In the PNET method, the following assumptions are used:

(1) Every two failure mechanisms having high correlation are considered to be perfectly correlated. If ρ_{ij}, as the coefficient of correlation between the ith and jth failure mechanisms, is larger than a prescribed critical value ρ_0, ρ_{ij} must be unity.

(2) Every two failure mechanisms having low correlation are considered to be statistically independent. If $\rho_{ij} < \rho_0$, ρ_{ij} must be zero.

It is to be noted that an appropriate value of ρ_0 should be selected in the implementation of the PNET method.

Using the technique proposed by Ishikawa and co-workers [9], significant failure mechanisms can be automatically found, according to the following procedure.

Step 1

The first representative failure mechanism is determined as the mechanism possessing the greatest failure probability or the least safety index. This can be done by solving the following mathematical programming problem:

Minimize safety index

$$\beta = \frac{\tilde{\phi}\mu_R - \mu_S}{\sqrt{(\tilde{\phi}\sigma_R)^2 + \sigma_S^2}} \tag{13}$$

subject to

$$\theta_j = \sum_{k=1}^{K} \bar{e}_k \cdot t_k \quad (j = 1, \ldots, J) \tag{14}$$

$$\sum_{k=1}^{k} \bar{e}_k \cdot t_k > 0 \tag{15}$$

where

$$\tilde{\phi}\mu_R = \sum_{j=1}^{J} \tilde{\phi}_j \bar{M}_{pj} \, |\theta_j| \tag{16}$$

$$\mu_S = \sum_{k=1}^{K} \bar{e}_k \cdot t_k \tag{17}$$

$$(\tilde{\phi}\sigma_R)^2 = \sum_{j=1}^{J} (\tilde{\phi}_j \bar{M}_{pj})^2 V_{mj}^2 \, |\theta_j|^2 \tag{18}$$

$$\sigma_S^2 = \sum_{k=1}^{K} \bar{e}_k^2 V_{ek}^2 t_k^2 \tag{19}$$

In this set of equations, $\bar{M}_{pj}$ is the mean value of the plastic moment at joint j, V_{mj} is the coefficient of variation of M_{pj}, $\bar{e}_k$ is the mean value of virtual work in the kth basic failure mechanism, V_{ek} is the coefficient of variation of e_k, t_k is a coefficient determining the combinational ratio of the basic mechanisms, j is the number of joints, and K is the number of basic failure mechanisms.

Step 2

The next step is to determine the second representative failure mechanism. Similarly, the second mode can be determined by minimizing, subject to the following constraint in addition to the constraints of Eqns. (14) and (15):

$$\tilde{\rho}(n) = \frac{\widetilde{\mathrm{Cov}}(Z_n, Z_1)}{\tilde{\sigma}_{Z_n} \cdot \tilde{\sigma}_{Z_1}} < \tilde{\rho}_0 \quad (n = 1, \ldots, N-1) \tag{20}$$

where Z_1 corresponds to the first failure mechanism.

$$\widetilde{\mathrm{Cov}}(Z_{n_1}, Z_1) = \sum_{j=1}^{J} \left| \sum_{k=1}^{K} t_{nk} C_{kj} \right| \cdot \left| \sum_{k=1}^{K} t_{1k} C_{kj} \right| (\tilde{\phi}_j \bar{M}_{pj})^2 V_{mj}^2 + \sum_{k=1}^{K} t_{nk} t_{1k} \bar{e}_k^2 V_{ek}^2 \tag{21}$$

$$\sigma_{Z_n}^2 = \sum_{j=1}^{J} (\tilde{\phi}_j M_{pj})^2 V_{mj}^2 \left| \sum_{k=1}^{K} t_{nk} C_{kj} \right|^2 + \sum_{k=1}^{K} t_{nk}^2 \bar{e}_k^2 V_{ek}^2 \tag{22}$$

$$\sigma_{Z_1}^2 = \sum_{j=1}^{J} (\tilde{\phi}_j \bar{M}_{pj})^2 V_{mj}^2 \left| \sum_{k=1}^{k} t_{1k} C_{kj} \right|^2 + \sum_{k=1}^{K} t_{1n}^2 \bar{e}_k^2 V_{ek}^2 \tag{23}$$

where C_{kj} is the virtual angle of rotation at joint j in the kth basic failure mechanism.

Step 3

Step 2 is repeated until $\tilde{P}_{fm}$ as the failure probability of the mth representative mechanism is sufficiently small in comparison to $\tilde{P}_{f1}$.

Step 4

Finally, the system failure probability $\tilde{P}_f$ is obtained as

$$\tilde{P}_f \cong \sum_n \tilde{P}_{fn} = \sum_n \Phi(-\tilde{\beta}_n) \tag{24}$$

As already mentioned, $\tilde{P}_f$ is represented here in terms of a fuzzy set. For two fuzzy numbers, it is sometimes difficult to determine which number is larger, because fuzzy numbers possess scatters around the central values [18]. Figure 2 is useful to understand this phenomenon.

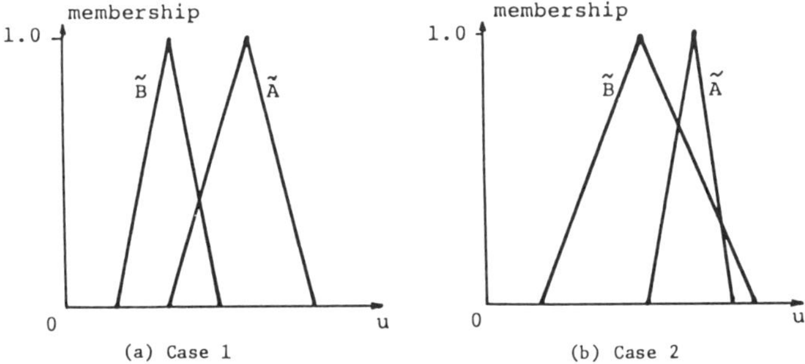

FIG. 2. Comparison of two fuzzy numbers.

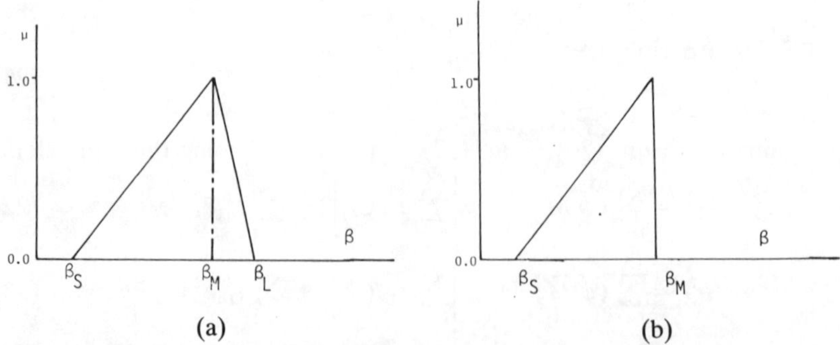

FIG. 3. Membership functions for the safety index. (a) Original function. (b) Approximate function.

Figure 2(a) shows that a fuzzy number A is clearly larger than the other fuzzy number B, while Fig. 2(b) shows that they have no definite difference. To order the fuzzy failure probabilities, an approximate method is used that considers the gravity center of the membership functions of failure probabilities. Failure probabilities less than 10^{-6} are not considered in the calculation, and failure mechanisms having failure probabilities with large right-hand side scatters are considered to be important. This method provides a result in which the ambiguity involved in the region of large values of failure probabilities is emphasized. For simplicity, it is assumed that the scatter of the left-hand side can be neglected. Thus, the membership function shown in Fig. 3(a) is reduced to the membership function shown in Fig. 3(b). Let us denote the minimum value, central value and maximum value of the fuzzy quantities by S, M and L, respectively.

Considering the gravity center of the membership function,

$$g = \frac{\int_0^1 x\mu(x)\,\mathrm{d}x}{\int_0^1 \mu(x)\,\mathrm{d}x} \tag{25}$$

the objective function expressed in Eqn. (20) results in

$$\frac{\beta_S + 2\beta_M}{3} \rightarrow \text{Min.} \tag{26}$$

Equation (26) implies that β_M is more important than β_S. On the other hand, the constraints regarding ρ_{ij} can be calculated by using the following ordering rule:

$$\rho_{ij} > \rho_0 \quad \text{if} \quad \rho_{ijS} + \rho_{ijM} + \rho_{ijL} > \rho_{0S} + \rho_{0M} + \rho_{0L} \tag{27}$$

In contrast to Eqn. (26), Eqn. (27) is derived by giving the equivalent weight to ρ_S, ρ_M and ρ_L.

NUMERICAL EXAMPLE

As an example, consider the portal frame whose statistical data are given in Table 1. Three damage patterns are considered, each of which is presented in Fig. 4. For this example, $\tilde{\phi} = (0{\cdot}75, 0{\cdot}25, 0{\cdot}25)$ is employed as a reduction factor. In Fig. 5, five basic independent failure mechanisms are presented, whose combinations provide all the possible failure mechanisms. In this simple example, the number of total failure mechanisms is eight. For each failure mechanism, safety indices are obtained as shown in Table 2, and by using Eqn. (3), the failure probability of each mechanism can be calculated corresponding to the safety indices obtained.

Using these values of $\tilde{P}_{fi}$ or $\tilde{\beta}_i$, the first representative failure mechanism is found as Mechanism 1, because it has the least safety index of $(\beta_{1S}, \beta_{1M}, \beta_{1L}) = (0{\cdot}29, 2{\cdot}33, 3{\cdot}88)$. The corresponding $\tilde{P}_{f1}$ is $(P_{f1S}, P_{f1M}, P_{f1L}) = (0{\cdot}53 \times 10^{-4}, 0{\cdot}96 \times 10^{-2}, 0{\cdot}38)$. Next, the second representative mechanism is chosen as the mechanism having the smallest value of $(\beta_S + 2\beta_M)$ among the failure mechanisms whose coefficients of correlation to the first mode are less than ρ_0. Their coefficients of correlation are obtained as

Mechanism 2: $\tilde{\rho}_{12} = (0{\cdot}568, 0{\cdot}810, 1{\cdot}336) \rightarrow 0{\cdot}925$
Mechanism 3: $\tilde{\rho}_{13} = (0{\cdot}534, 0{\cdot}750, 1{\cdot}058) \rightarrow 0{\cdot}781$
Mechanism 4: $\tilde{\rho}_{14} = (0{\cdot}0, 0{\cdot}0, 0{\cdot}0) \rightarrow 0{\cdot}0$

TABLE 1
Statistical data for the portal frame

	Mean value	*C.O.V.*
M_{pj} $(j = 1, \ldots, 8)$	40·0 N cm	0·2
H	0·5 N	0·2
P	1·0 N	0·2

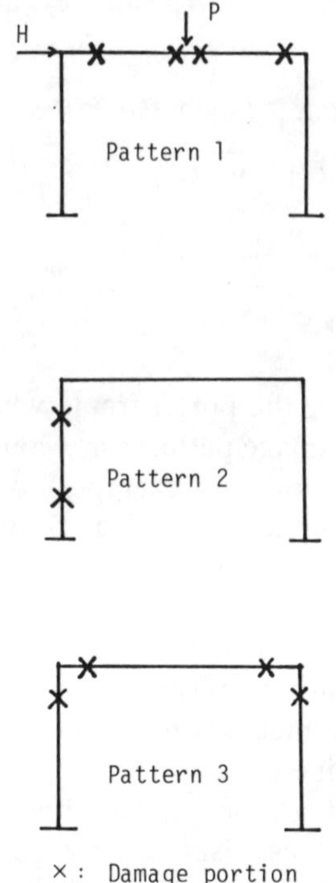

FIG. 4. Damage patterns for the portal frame.

Mechanism 5: $\tilde{\rho}_{15} = (0{\cdot}039, 0{\cdot}103, 0{\cdot}215 \rightarrow 0{\cdot}119$
Mechanism 6: $\tilde{\rho}_{16} = (0{\cdot}077, 0{\cdot}216, 0{\cdot}467) \rightarrow 0{\cdot}253$
Mechanism 7: $\tilde{\rho}_{17} = (0{\cdot}506, 0{\cdot}888, 1{\cdot}529) \rightarrow 0{\cdot}974$
Mechanism 8: $\tilde{\rho}_{18} = (0{\cdot}405, 0{\cdot}552, 0{\cdot}763) \rightarrow 0{\cdot}573$

From these results, Mechanism 8 is chosen as the second failure mechanism, because its value of $(\beta_S + 2\beta_M)$ is the greatest, and $\tilde{\rho}_{17}$ is less than 0·7, which is the prescribed critical value of $\tilde{\rho}_0$. Similarly, Mechanism 5 is chosen as the third representative mode. However, Mechanism 5 is not considered in the calculation of system failure probability, because its value is less than 10^{-6}. Finally, the system failure probability of the structure with the damage of pattern 1 is calculated as

$$\tilde{P}_f = \tilde{P}_{f1} + \tilde{P}_{f8} \tag{28}$$

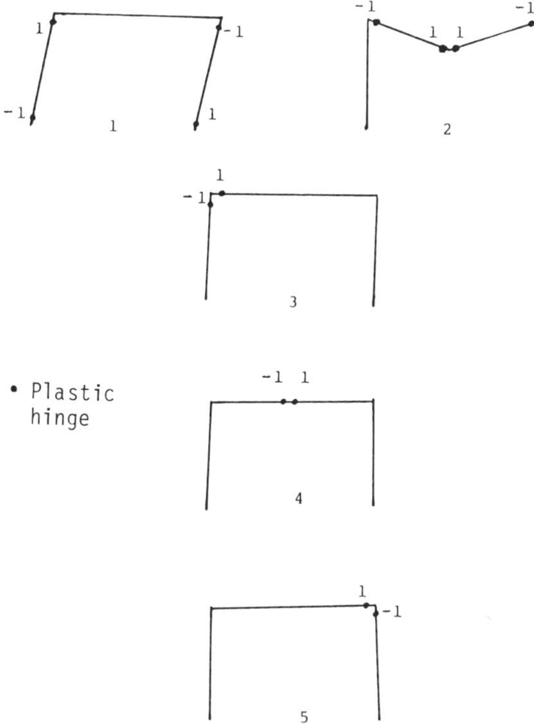

FIG. 5. Basic failure mechanisms.

TABLE 2
Safety indices obtained for damage pattern 1

	β_L	β_M	β_S	$(\beta_S + 2\beta_M)/3$
Z_1	3·88	2·34	0·29	1·66
Z_2	3·88	2·76	1·36	2·29
Z_3	3·88	3·15	2·29	2·86
Z_4	5·83	5·83	5·83	5·83
Z_5	5·83	5·52	5·13	5·39
Z_6	5·83	5·20	4·34	4·91
Z_7	3·97	2·90	1·49	2·43
Z_8	3·97	3·40	2·76	3·19

SUMMARY AND CONCLUSIONS

When a structure is damaged, its dominant failure modes may change because the grade, location and cause of damage can greatly affect the system reliability of the structure. However, an accurate estimation is very difficult due to technical and financial constraints, and it may be better to accept a vague or imprecise evaluation as a practical and meaningful basis for the reliability assessment of the damaged structure. For this purpose, fuzzy sets are useful because they can represent the verbal expression of experienced engineers in a clear and informative manner.

In this analysis, the PNET method has been extensively applied to calculate the system failure probability of a damaged redundant structure. Since the failure probabilities and the coefficients of correlation were calculated as fuzzy quantities, some modification was necessary for the calculation process for the system failure probability. Using the technique proposed by Ishikawa and co-workers, it is possible to identify automatically the significant failure mechanisms which are sufficient to calculate the system failure probability with accuracy. An approximate calculating method has also been proposed to make the calculation of fuzzy quantities simpler.

REFERENCES

[1] M. R. Gorman and F. Moses, *Proceedings of the Symposium on Probabilistic Methods in Structural Engineering,* ASCE, St. Louis, 251 (1981).
[2] D. M. Frangopol, Effects of redundancy deterioration on the reliability of truss systems and bridges, ASCE Convention, Seattle, WA (1985).
[3] J. T. P. Yao, *Soil Dynamics Earthquake Eng.*, **1**(3), 130 (1982).
[4] N. Shiraishi, H. Furuta and M. Sugimoto, *Proc. ICOSSAR-4,* Kobe, Japan, IASSAR, I505 (1985).
[5] N. Shiraishi, and H. Furuta, *Proceedings of the NSF Workshop on Civil Engineering Applications of Fuzzy Sets,* Purdue University, West Lafayette, USA, 193 (1985).
[6] J. T. P. Yao, *J. Eng. Struct.*, **1,** 245 (1979).
[7] L. A. Zadeh, *Information and Control,* **8,** 338 (1965).
[8] A. H.-S. Ang and H.-F. Ma, *Proc. ICOSSAR-3,* Trondheim, Norway, Elsevier, Amsterdam, 295 (1981).
[9] T. Mihara, M. Iizuka, N. Ishikawa and K. Furukawa, *J. Struct. Eng., JSCE,* **32A,** 475 (1986) (in Japanese).
[10] J. T. P. Yao, *Safety and Reliability of Existing Structures,* Pitman Advanced Publishing Program, Boston, MA (1985).

[11] J. T. P. Yao and H. Furuta, *J. Probabilistic Mechanics,* **1**(1), 58 (1986).
[12] L. A. Zadeh, *Information Science,* **8,** 99 (1975).
[13] D. I. Blockley, *The Nature of Structural Design and Safety,* Ellis Horwood, Chichester, U.K., (1980).
[14] C. B. Brown, H. Furuta, N. Shiraishi and J. T. P. Yao, *Proceedings of the 1st International Conference on Fuzzy Information Processing,* Kauai, Hawaii (1987).
[15] D. Dubois and H. Prade, *Fuzzy Sets and Systems: Theory and Applications,* Academic Press, New York, (1980).
[16] D. Dubois and H. Prade, *Fuzzy Sets and Systems,* **2,** 327 (1979).
[17] Y. Murotsu, H. Okada and S. Matsuzaki, *Proc. ICOSSAR-4,* Kobe, Japan, IASSAR, I117 (1985).
[18] J. Buckley, *Fuzzy Sets and Systems,* **15,** 1 (1985).

Reliability Analysis of Bridge Structures Using a Simulation Method

NOBUYOSHI TAKAOKA, WATARU SHIRAKI and SHIGEYUKI MATSUHO

Department of Civil Engineering, Tottori University, Tottori-shi, Japan

ABSTRACT

After reviewing past work on the safety and reliability of bridges, a method for the reliability analysis of highway bridges subjected to random vehicular loads is demonstrated. By simulation analysis, the bending moment induced in main girders is assumed to follow the type I extreme value distribution law. The joint probability density function of the bending moment and its derivative with respect to the abscissa along the bridge axis are evaluated by the method of translation processes. Employing the theory of level-crossings by random processes, the reliability of the main girders in a space domain as well as in a time domain are evaluated. The randomness of vehicular loads in the direction of the bridge axis as well as in the direction perpendicular to this axis are considered.

INTRODUCTION

The loading conditions of highway bridges in Japan have become more severe from year to year due to the rapid increase in the number and tonnage of motor vehicles crossing the bridges. As a result, maintenance work on highway bridges has become one of the most important tasks for the Highway Authorities, whose duty is always to keep the roadway and the traffic flow in a normal and satisfactory state.

Since 1979, we have been carrying out a reliability analysis of highway bridges using the theory of random processes and a simulation technique, and the first result was reported in 1982 [1, 2].

To grasp the reality of these increasingly severe loading conditions, an extensive survey on actual traffic streams was conducted in 1983 by the Hanshin (Osaka–Kobe) Expressway Public Corporation, and the results obtained have been reported elsewhere [3–5].

Using the data from this survey, we have been carrying out since then theoretical research on the following items from the statistical and probabilistic point of view:

(1) probabilistic modeling of vehicular loads by a simulation method [6–8];
(2) reliability analysis of highway bridges in the space domain as subjected to simulated vehicular loads using the theory of random processes [8];
(3) reliability analysis of highway bridges in the space domain and in the time domain [9, 10], with application of the method of translation processes [11];.
(4) reliability analysis of the concrete slab roadway of a bridge by the simulation method [12];
(5) reliability analysis of rigid-frame pier structures supporting highway bridges [13], involving the load-combination problem [14].

We now report mainly on item (3), supplementing the earlier results with those recently obtained.

In our early studies, we tentatively assumed that the bending moment on a bridge girder $\tilde{M}(x)$ was a Gaussian random distribution along the bridge span. The tilde mark ($\sim$) over a letter indicates that the letter represents a random quantity. The reliability of the bridge girder was evaluated using the theory of level-crossings by random processes. It was made clear by these studies that:

(1) the bending moment $\tilde{M}(x)$ obeys the type I extreme value distribution (Gumbel distribution) law rather than the Gaussian law;
(2) for better modeling of vehicular loads, their randomness in the direction of the bridge span as well as in the direction perpendicular to it should be considered, since highway bridges usually have multiple lanes. In the previous studies, the randomness of vehicular loads only in the direction of the bridge span was considered.

We will first show a reliability analysis of highway bridges when the bending moment along the girder is of a non-Gaussian distribution, a numerical example being presented of a model bridge with a two-span continuous girder (single lane).

Next, we will model the vehicular loads in a fully congested traffic stream state, taking the randomness in both the span and width directions. This vehicular load model will be compared with the load model obtained by assuming randomness only in the bridge span direction. To model the vehicular loads, both the simulation technique and theory of random processes are employed, incorporating into the simulation process the actual data on traffic streams provided by the Hanshin Expressway Public Corporation. Using the vehicular load model thus obtained, the reliability of multi-girder bridges (multiple lanes) will be analyzed.

RELIABILITY ANALYSIS OF HIGHWAY BRIDGES USING TRANSLATION PROCESSES

Evaluation of Joint Probability Density Function

To apply the theory of level-crossings by random processes, say $\tilde{M}(x)$, to reliability analysis, we need to know about the joint distribution of the random process $\tilde{M}_1(x) = \tilde{M}(x)$ and its derivative $\tilde{M}_2(x) = \mathrm{d}\tilde{M}(x)/\mathrm{d}x$. The joint probability density function $f_{M_1M_2}(M_1, M_2 \mid x)$ is little known for the general case when $\tilde{M}(x)$ is of a non-Gaussian distribution, but this difficulty may be overcome by making use of the method of translation processes proposed by Grigoriu [15].

Figure 1 shows the relationship between the standard Gaussian variable $\tilde{M}_N(x)$ with a probability distribution function $\Phi(x)$ and the non-Gaussian variable $\tilde{M}(x)$ with a distribution function $F(x)$. Note that two random processes $\tilde{M}_N(x)$ and $\tilde{M}(x)$ are both considered random variables if their argument x is assumed to be fixed.

Now, referring to Fig. 1 and with a common probability P for $\tilde{M}_N(x)$ and $\tilde{M}(x)$, we can deduce that the following relationship holds:

$$\begin{aligned} P &= F\{M(x)\} = \Phi\{M_N(X)\} \\ \tilde{M}(x) &= (F^{-1} \circ \Phi)\{\tilde{M}_N(X)\} = g\{\tilde{M}_N(x)\} \end{aligned} \tag{1}$$

in which F^{-1} = the inverse function of the distribution function F, and $g = (F^{-1} \circ \Phi)$ = the composite function of F^{-1} and Φ.

According to Grigoriu [15], the joint probability density function

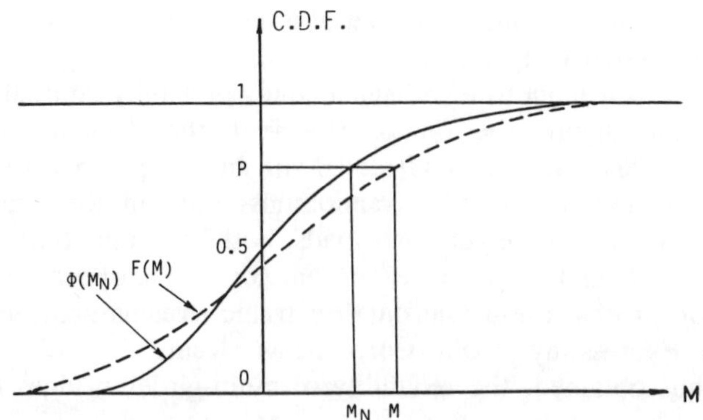

FIG. 1. Relationship between standard Gaussian variable and non-Gaussian variable.

$f_{M_1M_2}(M_1, M_2 \mid x)$ for a non-Gaussian random process $\tilde{M}_1(x) = \tilde{M}(x)$ and its derivative $\tilde{M}_2(x) = \mathrm{d}\tilde{M}(x)/\mathrm{d}x$ for a fixed value of their argument x is given by

$$
\begin{aligned}
f_{M_1M_2}(M_1, M_2 \mid x) = {} & \frac{f_{M_1}(M_1 \mid x) f_{M_2}(M_2 \mid x)}{\sqrt{1-\rho_N^2(x)}} \\
& \times \exp\Bigg[-\frac{\rho_N(x)}{2\{1-\rho_N^2(x)\}} (\rho_N(x)[(g_{M_1}^{-1}\{M_1(x)\})^2 \\
& + (g_{M_2}^{-1}\{M_2(x)\})^2] - 2g_{M_1}^{-1}\{M_1(x)\} g_{M_2}^{-1}\{M_2(x)\}) \Bigg]
\end{aligned}
\tag{2}
$$

in which $M_{N_1}(x) = g_{M_1}^{-1}\{M_1(x)\}$, $M_{N_2}(x) = g_{M_2}^{-1}\{M_2(x)\}$ and $\rho_N(x) =$ the correlation coefficient of two standard Gaussian variables $\tilde{M}_{N_1}(x)$ and $\tilde{M}_{N_2}(x)$ at any fixed value of the argument x.

Reliability Evaluation of the Main Girder

First of all, it should be noted that the safety of a bridge is decisively governed by vehicular loads which fully occupy the entire span of the bridge in a fully congested traffic stream state. Such traffic jams will occur repeatedly in the space domain (i.e. over the entire bridge span) as well as in the time domain (i.e. during the lifetime of the bridge). In the state of traffic congestion, motor vehicles move very slowly or come to a complete standstill. Therefore, the dynamic effect due to moving vehicles has not been considered.

Reliability Analysis in the Space Domain

The event in which the external bending moment $\tilde{M}(x)$ exceeds a given allowable level M_a over the entire span length l of the bridge may be assumed to be very rare. Therefore, the probability Q of the event that this excess load ($|\tilde{M}(x)| > M_a$) takes place at least once over the entire span length l of the bridge is given by

$$Q = 1 - \exp(-\mu) \tag{3}$$

in which μ = the average rate of the excess load already mentioned. We define the failure of the girder by this excess.

The average rate of excess μ is given by

$$\mu = \int_0^l \{p_+(0 \mid x) + p_-(0 \mid x)\}\, dx \tag{4}$$

in which

$$\begin{aligned} p_+(0 \mid x) &= \int_0^{\infty} f_{Z_1\dot{Z}_1}(0, \dot{Z}_1 \mid x)\dot{Z}_1\, d\dot{Z}_1 \\ p_-(0 \mid x) &= \int_{-\infty}^{0} f_{Z_2\dot{Z}_2}(0, \dot{Z}_2 \mid x)\dot{Z}_2\, d\dot{Z}_2 \end{aligned} \tag{5}$$

and

$$\tilde{Z}_1(x) = \tilde{M}(x) - M_a; \qquad \tilde{Z}_2(x) = \tilde{M}(X) + M_a \tag{6}$$

The joint probability density functions for the non-Gaussian processes, $\tilde{Z}_1(x)$ and $\tilde{Z}_2(x)$, which we need to evaluate Eqn. (5), can be derived by making use of Eqn. (2). It can be shown that $\rho_N(x)$ in Eqn. (2) may be approximated by the correlation coefficient $\rho(x)$ between the bending moment $\tilde{M}(x)$ and the shearing force $\tilde{Q}(x)$ at the same section x; that is

$$\rho_N(x) = \rho(x) \tag{7}$$

Reliability Analysis in the Time Domain

Let the probability distribution function of the bending moment $\tilde{M}(x)$ be designated by $F_M(m)$, given once a traffic jam has occurred. Then, the (failure) probability Q^* of the largest bending moment after the occurrence of N traffic jams exceeding in the time domain an allowable level M_a is given by (the power law)

$$Q^* = 1 - \{F_M(M_a)\}^N \tag{8}$$

The probability distribution function $F_M(m)$ can be derived from Eqn. (3); that is

$$\begin{aligned} F_M(m) &= \text{Prob}[|\tilde{M}(x)| < m] \\ &= 1 - Q = \exp(-\mu) \\ &= \exp\left[-\int_0^l \{p_+(0 \mid x) + p_-(0 \mid x)\} \, \mathrm{d}x \right] \end{aligned} \tag{9}$$

Numerical Example

We will demonstrate here the gist of a reliability analysis of a model highway bridge with a two-span continuous girder of total length $l = l_1 + l_2 = 100$ m. The roadway is assumed to be supported by a single girder and subjected to a simulated vehicular load $\tilde{q}(x)$, which is randomly distributed over the entire span of the bridge (see Fig. 2(a)). The probabilistic characteristics of the distributed load $\tilde{q}(x)$ are [6, 7]:

the expected value:

$$\bar{q} = 0{\cdot}4586 \text{ t m}^{-1} \fallingdotseq 4{\cdot}586 \text{ kN m}^{-1},$$

the variance:

$$D_q = 0{\cdot}4064 \text{ (t m}^{-1})^2 \fallingdotseq 40{\cdot}64 \text{ (kN m}^{-1})^2$$

the autocorrelation function:

$$K_q(\tau) = D_q \cdot \exp(-\alpha\tau) \quad \text{with} \quad \alpha = 0{\cdot}3164 \text{ m} \quad \text{and} \quad \tau = x_2 - x_1.$$

Based on statistical processing of simulated values of the bending moment $\tilde{M}(x)$ induced by the random load $\tilde{q}(x)$, it can be assumed that $\tilde{M}(x)$ follows the type I extreme value distribution (see also Fig. 5).

Figure 3 shows the relationship between the allowable bending moment level M_a and the span ratio $n = l_2/l_1$, with a parameter Q which is the specified (target) value of the failure probability. For comparison, a diagram of the same kind is shown in Fig. 4 for the case when the bending moment is assumed to follow a Gaussian distribution.

Next, we turn to reliability analysis in the time domain. For this purpose, we must have the distribution function $F_M(m)$ of the bending moment $\tilde{M}(x)$, the exact expression of which is given by Eqn. (9). However, we will resort here to an approximate solution which can be

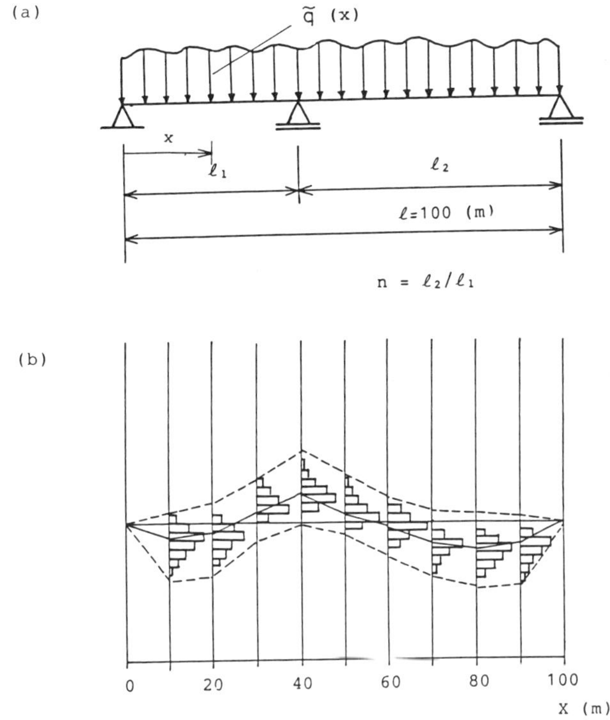

FIG. 2. Model bridge with two-span continuous girder. (a) Randomly distributed load. (b) Frequency distribution of bending moment at several cross-sections.

obtained by plotting the values of $(1-Q)$ versus the allowable moment level M_a on probability paper. This procedure was done for a span ratio of $n = 1{\cdot}6$ on Gumbel probability paper (see Fig. 5). From this diagram, we can find that the bending moment $\tilde{M}(x)$ follows the type I extreme value distribution. For other values of n, we can also obtain the same conclusion. Thus, we have the expression

$$F_M(m) = \exp[-\exp\{-\alpha^*(m-u)\}] \tag{10}$$

in which $\alpha^* = 0{\cdot}0176\,(\mathrm{t\,m})^{-1} \fallingdotseq 0{\cdot}00176\,(\mathrm{kN\,m})^{-1}$ and $u = 183{\cdot}2$ $(\mathrm{t\,m}) \fallingdotseq 1832\,(\mathrm{kN\,m})$. The values of these parameters were estimated by means of regression analysis.

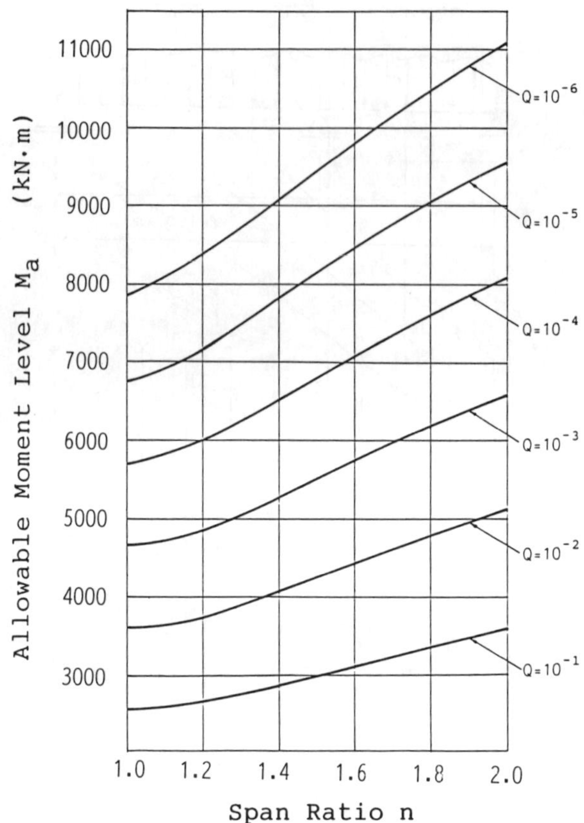

FIG. 3. Allowable moment level versus span ratio for a specified failure probability (type I extreme value distribution).

Figure 6 shows the allowable moment level M_a versus the number of traffic congestion events N for several values of the specified failure probability Q^*. In this diagram, the dashed line (L-20i) shows the design moment level (with the impact effect included) that was calculated according to the current Japanese Design Specifications for Highway Bridges (1980). In the actual design process, this design value is multiplied by a safety factor of S.F. = 1·7 (see the curve marked with the notation L-20i × 1·7 in Fig. 6). We can find from this diagram that the current Japanese Design Specifications anticipate a failure probability of say $Q^* = 10^{-3}$.

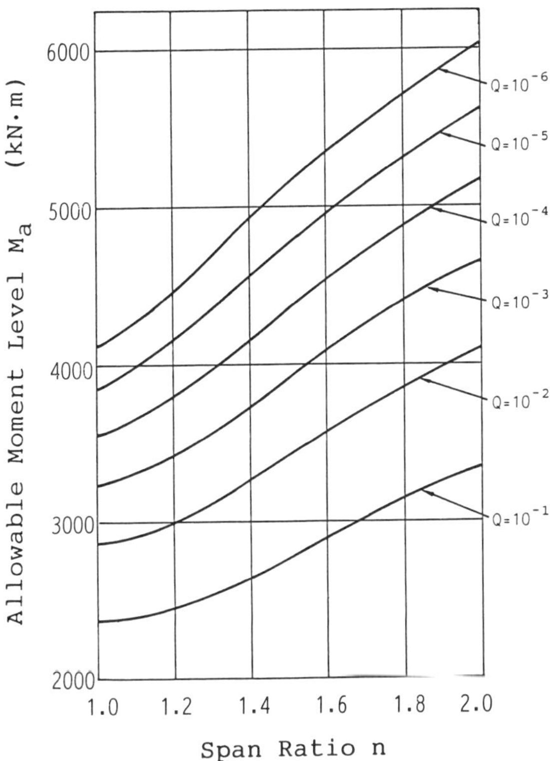

FIG. 4. Allowable moment level versus span ratio for a specified failure probability (Gaussian distribution).

Assuming the number of traffic congestion events in 50 years to be $N = 5000$, the relationship between the allowable moment level M_a and the span ratio n for several values of Q^* as a parameter is indicated in Fig. 7. For comparison, a diagram of the same kind is shown in Fig. 8, which was drawn for the case when the bending moment was assumed to follow a Gaussian distribution.

A comparison between Fig. 7 and Fig. 3 reveals that the consideration of the randomness of vehicular loads in the space domain as well as in the time domain requires much higher values of the allowable moment level M_a than when considering the randomness in the space domain only.

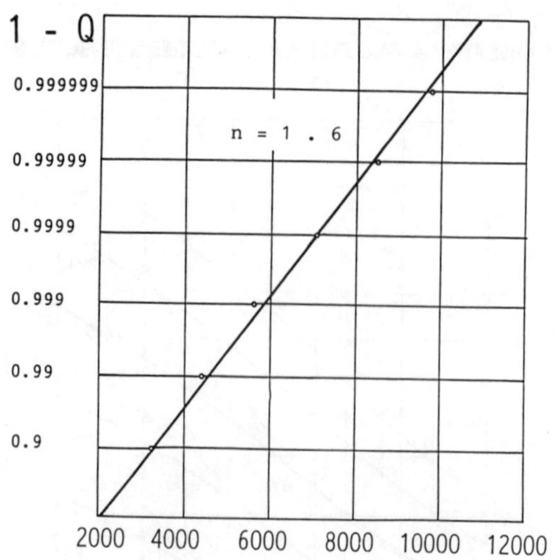

FIG. 5. Fitness of bending moment to a type I extreme value distribution (on Gumbel probability paper).

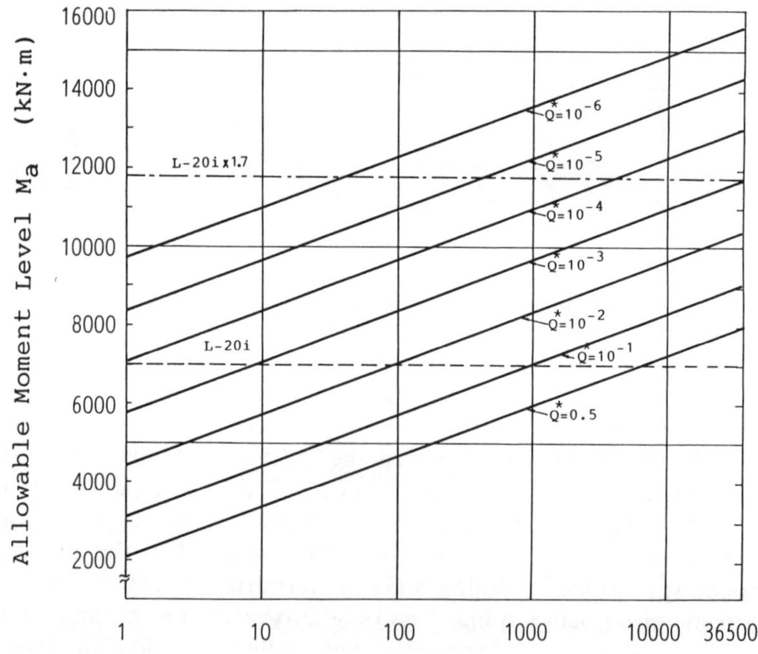

FIG. 6. Relationship between allowable moment level and number of traffic congestion events for a specified failure probability.

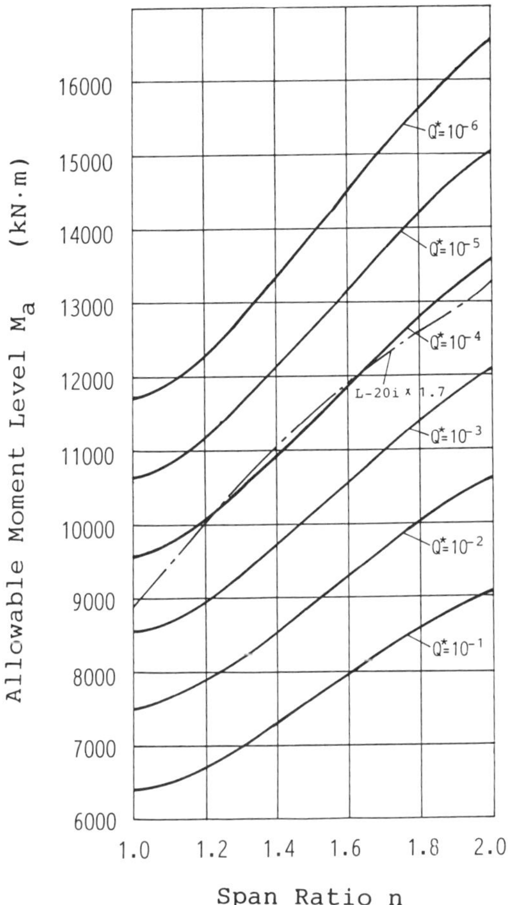

FIG. 7. Allowable moment level versus span ratio for a specified failure probability with vehicular randomness in two directions (type I extreme value distribution).

RELIABILITY ANALYSIS OF MAIN GIRDERS BY CONSIDERING RANDOMNESS OF VEHICULAR LOADS IN TWO ORTHOGONAL DIRECTIONS

In the section concerning the reliability evaluation of the main girder and in the numerical example, we considered the randomness of

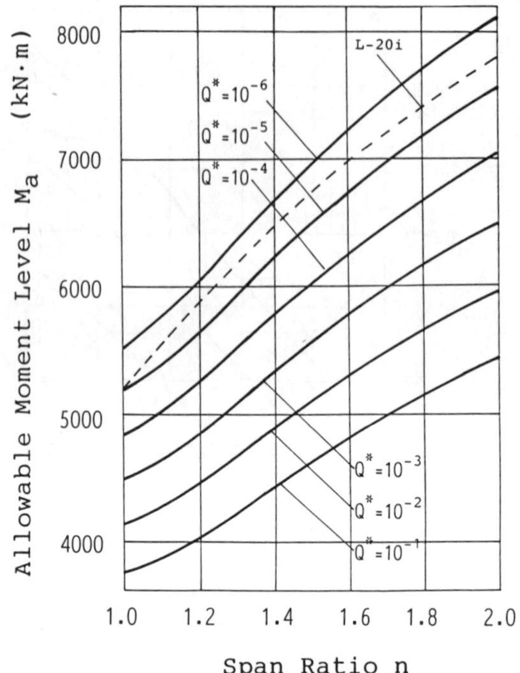

FIG. 8. Allowable moment level versus span ratio for a specified failure probability with vehicular randomness in two directions (Gaussian distribution).

vehicular loads in the direction of the bridge span only. We now consider in addition the randomness of vehicular loads in a direction perpendicular to the bridge span. For simplicity, the former (bridge span direction) is called the 'longitudinal direction', and the latter the 'lateral direction'.

Figure 9(a) shows schematically how to model the vehicular load, the total weight of a vehicle being divided into two parts, each of which is uniformly distributed.

As an example, we will consider a model bridge with a two-lane concrete slab roadway which is supported by five main girders. The modeled vehicular loads are applied on this roadway as shown schematically in Fig. 9(b), the two lanes being in the state of full congestion. In considering randomness in the lateral direction of vehicular loads, the position of the vehicles is also assumed to follow a

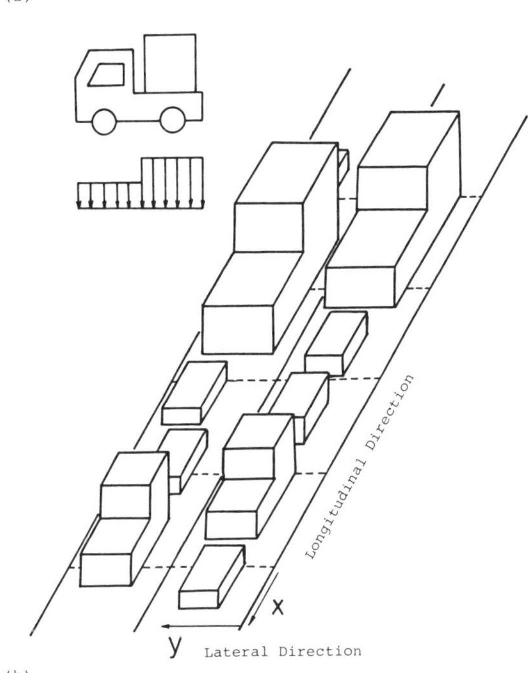

FIG. 9. Model of live vehicular loads. (a) Distribution of vehicular weight. (b) Roadway subjected to fully congested vehicular loads.

certain probability distribution law, which was numerically estimated based on field observation results and incorporated into the simulation process. The width B of vehicles is a deterministic variable which varies according to the type of vehicle.

Referring to the influence functions of each girder (see Fig. 10(b)), the value of the vehicular load sustained by each girder is determined and the bending moment $\tilde{M}_i(x)$ induced in the ith girder by the vehicular load is calculated. The reliability analysis proceeds in a similar manner to that given above when considering the main girder. The bending moment $\tilde{M}_i(x)$ is also assumed here to follow the type I extreme value distribution law, and the reliability of each girder in the space domain and time domain can be examined.

The result is summarized in column 3 of Table 1, in which the

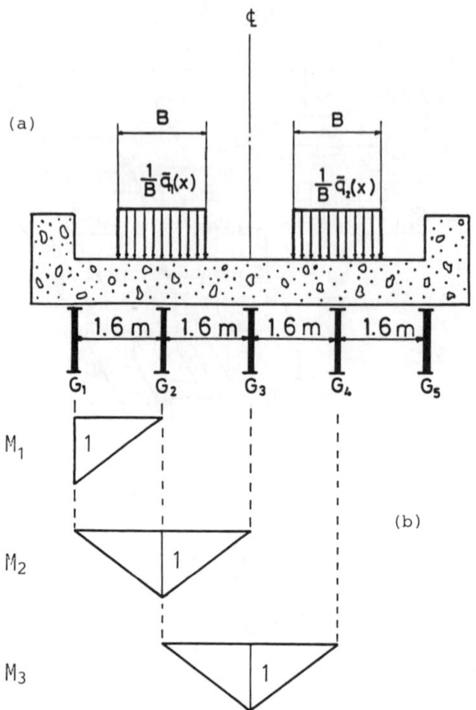

FIG. 10. Cross-section of a model bridge. (a) Concrete slab roadway supported by 5 girders. (b) Influence functions.

allowable moment level M_a was determined for several values of specified failure probability Q^* with the total number of congestion events during the lifetime $N = 5000$. For comparison, the result for the case in which only randomness in the longitudinal direction was considered is summarized in column 4 of Table 1. In the latter case, the width B of vehicles was assumed to be a deterministic value, which is equal to the mean of all B values for each type of vehicle, weighted according to the vehicle composition of the traffic. In addition, the vehicles were assumed to be driven in the center of each lane.

Table 1 demonstrates clearly that considering the randomness of vehicular loads in both directions produces markedly different values for the allowable moment level from those obtained by considering randomness in the longitudinal direction only.

TABLE 1
Allowable moment level M_a (in kN m)

Girder no.	q^*	*Ma*	
		Two directions	*One direction*
G_1	0·5	1395	807·6
	10^{-1}	1630	934·5
	10^{-2}	1923	1093
	10^{-3}	2211	1248
	10^{-4}	2499	14£3
G_2	0·5	4684	5330
	10^{-1}	5412	6167
	10^{-2}	6320	7212
	10^{-3}	7212	8237
	10^{-4}	8102	9261
G_3	0·5	2101	1245
	10^{-1}	2433	1424
	10^{-2}	2846	1647
	10^{-3}	3252	1866
	10^{-4}	3657	2084

CONCLUDING REMARKS

Several aspects of reliability analysis for highway bridges have been reported in this paper, after summarizing the topics of investigation. A method of reliability analysis for highway bridges using the theory of random processes and a simulation technique indicated that the bending moment induced in the main girders of a bridge by live vehicular loads followed the type I extreme value distribution. Using the method of translation processes, the joint probability density of the bending moment and its derivative with respect to the abscissa (i.e. shearing force) was obtained to evaluate the upcrossing rate of the bending moment. Numerical examples were presented to illustrate the analysis methods used.

REFERENCES

[1] N. Takaoka, Reliability analysis of structures using theory of random processes and simulation method, in *Analysis of Random Capacity of Structures,* collected papers, J. Murzewski, Ed., Polish Academy of Sciences, 89 (1982–6).

[2] N. Takaoka, Reliability analysis of highway bridges subjected to vehicular loads using theory of random processes and simulation method, *Proc. EUROMECH 155, June 15–17, 1982, Lyngby, DIALOG 6-82,* Danish Engineering Academy, Lyngby, Denmark, 185 (1982).
[3] HDL Committee, Survey and measurement of traffic loads on Hanshin Expressway network and their analysis, in *Report of Subcommittee on Live Loads* (*LLS*), Hanshin Expressway Public Corporation (1984) (in Japanese).
[4] HDL Committee, Investigation of design loads system on Hanshin Expressway network, in *Report of HDL Committee,* Hanshin Expressway Public Corporation (1986) (in Japanese).
[5] I. Konishi, H. Kameda and T. Matsumoto, Traffic load measurement and probabilistic modeling for structural design of urban expressway, *Proc. ICOSSAR '85,* International Association for Structural Safety and Reliability, Kyoto, Japan, III.141 (1985).
[6] S. Matsuho, W. Shiraki, N. Takaoka and K. Yamamoto, A probabilistic evaluation of vehicular loads, *Proc. ICOSSAR '85,* International Association for Structural Safety and Reliability, Kyoto, Japan, I.490 (1985).
[7] W. Shiraki, N. Takaoka, S. Matsuho and K. Yamamoto, Modeling of vehicular loads in a fully congested traffic stream state on urban expressway bridges, *Proc. JSCE,* No. 362/I–4, 239 (1985) (in Japanese).
[8] N. Takaoka, W. Shiraki and S. Matsuho, Reliability analysis of highway bridge subjected to vehicular loads using theory of random functions, *Proc. JSCE,* No. 334, 79 (1983) (in Japanese).
[9] W. Shiraki, N. Takaoka, S. Matsuho and K. Yamamoto, Reliability analysis of various types of girder bridges on urban expressway network using theory of random processes and simulation method, *Proc. ICOSSAR '85,* International Association for Structural Safety and Reliability, Kyoto, Japan, III.185 (1985).
[10] W. Shiraki, S. Matsuho, K. Yamamoto and N. Takaoka, Reliability analysis of a continuous two-span bridge girder, *Proc. JSCE,* No. 368/I–5, 225 (1986) (in Japanese).
[11] W. Shiraki, S. Matsuho and N. Takaoka, Reliability analysis of highway bridges using translation processes, *J. Struct. Eng.* **32A**, 561 (1986) (in Japanese).
[12] W. Shiraki, S. Matsuho and N. Takaoka, Reliability analysis of highway bridge slab using simulation method, *J. Struct. Eng.*, **32A,** 571 (1986) (in Japanese).
[13] W. Shiraki, N. Takaoka, S. Matsuho and K. Yamamoto, Reliability analysis of rigid-frame piers on urban expressway bridges, *J. Struct. Eng.*, **31A,** 313 (1985) (in Japanese).
[14] W. Shiraki, S. Matsuho and N. Takaoka, Load combination analysis and reliability analysis of steel rigid-frame piers supporting bridges constructed on urban expressway network, *Proc. ICASP5,* Vancouver, Canada, Vol. 1, Institute for Risk Research, University of Waterloo, Waterloo, Canada, 206 (1987).
[15] M. Grigoriu, Crossing of non-Gaussian translation process, *J. Eng. Mech. Div., ASCE,* **110**(ST4), 610 (1984).

Application of the Markov Chain to the Reliability Analysis of Fatigue Crack Propagation with Non-destructive Inspection

YOSHIHIRO SHIMADA
Olympus Optical Co., 2951, Ishikawa, Hachiooji, Japan.

TAKAO NAKAGAWA and HISANOBU TOKUNO
Faculty of Engineering, Kobe University, Rokkodai, Nada, Kobe, Japan

ABSTRACT

Even in a structure manufactured under complete quality control, it is impossible to exclude some defects which initiate cracks in the early stage of their lives. Consequently, the non-destructive inspection (NDI) is ordinarily performed on the structures which require safety.

In the present paper, by applying the initial crack distribution and the stochastic matrix (developed in a previous paper) to the crack propagation process during NDI, the probability of failure was calculated in both cases of 'repair model' and 'replacement model'. Then, the reliabilistic analysis was performed for a structure subjected to NDI, taken as a model of fracture. The effect of NDI as well as that of the variability in crack propagation on the probability of failure were also studied.

INTRODUCTION

The strength of an original material has statistically variable properties. Therefore, the behavior of crack propagation of structural members made of this material due to external loadings has essentially stochastic properties, and the process of crack propagation is also considered as stochastic. There have been many studies on fatigue crack propagation and the detailed results have already been published, but very few treated the crack length or its state as the random variable [1–4].

Bogdanoff *et al.* considered the process of fatigue crack propagation as a kind of Markov process, gave the initial statistical distribution of crack length, and calculated the mean, variance and distribution of crack length at an arbitrary time, as well as the number of cycles to reach the specified length of crack (that is, the crack propagation life) by applying the theory of the Markov chain (a discrete Markov process with the stochastic transition matrix) [3, 4]. The authors have performed a similar study to those just mentioned, indicating that fatigue crack propagation is a process having the constant ratio of $r = p/q$ (where p and q are the elements in the stochastic transition matrix of the Markov chain) and verifying the possibilities of applying the Markov chain to fatigue crack propagation [5].

The present study extends the previous study [5], applies this theory to the case of non-destructive inspection (NDI), and investigates the probability of a fracture vs. time relationship and the effect of the number of inspections on this relationship [6].

Even with a structure made under careful quality control, it is impossible to completely exclude defects that can generate cracks in the early stages of the service period. Consequently, in any structure requiring a high level of safety, NDI is carried out. The detectability properties of NDI presumably have a stochastic nature; therefore, it is necessary to use probabilistic or statistical techniques to assess the effect of NDI [7, 8]. Kitagawa *et al.* systematically developed Shinozuka's method of using the conditional probability in reliability analysis considering NDI, and introduced an equation to obtain the probability of fracture in both cases of the 'the repair model' and 'the replacement model' [9, 10]. Bogdanoff introduced the expression for the probability of fracture in the case with NDI by extending his theory already described on the discrete Markov process and using a stochastic transition matrix in the case of 'the repair model', but he did not refer to 'the replacement model'.

In the present paper, an equation to find the probability of fracture is shown by using the stochastic transition matrix in both cases of 'the repair model' and the 'replacement model'. Reliability analyses of structures subjected to NDI are performed, and the effect of NDI and of the variability of crack propagation on the probability of fracture are studied.

THEORY

Fracture Model

A steel structure is assumed, and a surface crack of semi-circular form with radius a_0 is considered as shown in Fig. 1, the crack being

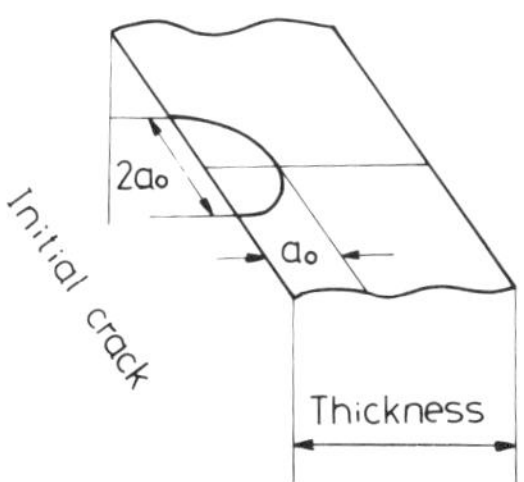

FIG. 1. Fracture model.

assumed to extend while holding a similar form. The fracture is judged to occur when the value of the crack length a reaches the value of the thickness of the plate t, that is $a = t$.

Stochastic Model

Applying the theory of the Markov chain used in a previous paper [5], the process of crack propagation is modelled by the stochastic transition matrix. As the progress of the system is specified stochastically by using this matrix, it is a feature of this theory not to use any relationship specifying the central tendency of the crack length, as is used in the Paris–Erdogan equation, etc.

Application to Non-destructive Inspection (NDI)

The Case Accompanied by an In-service Inspection (ISI)

The probability of detecting cracks is represented by

$$D_j = \text{Prob}[\text{detecting crack} \mid \text{crack length } a_j] \tag{1}$$

If cracks are not detected until they reach a crack length a_{i-1}, and can first be detected when reaching a_i, D_j can be expressed by Eqn. (2).

$$\begin{aligned} D_j &= 0 \qquad (j = 1, 2, \ldots, i-1) \\ 0 \le D_j &\le 1 \qquad (j = i, i+1, \ldots, b-1) \end{aligned} \tag{2}$$

where b represents the critical crack state. This case will be developed by considering the two cases of the 'repair model' and the 'replacement model'.

Repair model. In this model, there appears to be no crack after inspection and repair. The initial statistical distribution of the crack state is given by

$$\mathbf{P}_0 = \{p_0(1), p_0(2), \ldots, p_0(b)\} \tag{3}$$

where subscript 0 indicates the initial time and the numerals $1, 2, \ldots, b$ in parentheses show the state of the crack length. Next, using the Markov chain as a stochastic model, the stochastic transition matrix is given by

$$\mathbf{P} = \begin{pmatrix} p_1 & q_1 & 0 & \cdots & 0 & 0 \\ 0 & p_2 & q_2 & \cdots & 0 & 0 \\ \vdots & \vdots & \vdots & & \vdots & \vdots \\ 0 & 0 & 0 & \cdots & p_{b-1} & q_{b-1} \\ 0 & 0 & 0 & \cdots & 0 & 1 \end{pmatrix} \tag{4}$$

where p_j is the probability of maintaining the crack length a_j, q_j is the probability of increasing from a_j, to a_{j+1}, and $p_j + q_j = 1$. (The state of cracking is a kind of damage state and it is considered as the accumulation of the unit damage due to each stress reversal. As the crack length a_j corresponds to the cumulative damage d_j and the progress of d_j is cumulative, this process is monotonically increasing and never regresses; that is, there is no change such as $j \to j-1$.)

From the theory of the Markov chain [11], the probability vector of the crack state at the time x_1 is

$$\mathbf{P}_{x1} = \{p_{x1}(1), p_{x1}(2), \ldots, p_{x1}(b)\} = \mathbf{P}_0\mathbf{P}^{x1} \tag{5}$$

The probability vector of the crack state immediately after inspection and repair at the time x_1 is

$$\mathbf{P}_{01} = \{p_{01}(1), p_{01}(2), \ldots, p_{01}(b)\} \tag{6}$$

(The 0 in subscript 01 represents the initial distribution, and the 1 indicates the in-service inspection.) As this model is assumed to revert to the no-crack state after any crack has been discovered and this portion repaired, the product of the probability of the existence of a crack length a_j and the probability of not detecting cracks of the length a_j, that is $1 - D_j$ at the time 01, gives the initial distribution of crack length at that time. Therefore,

$$p_{01}(j) = (1 - D_j)p_{x1}(j) \tag{7}$$

After the inspection at the time x, it follows that

$$\mathbf{P}_x = \{p_x(1), p_x(2), \ldots, p_x(b)\} = \mathbf{P}_{01}\mathbf{P}^{x-x1} \tag{8}$$

The probability of fracture at the time x is

$$P_f = p_x(b) \tag{9}$$

and the hazard rate (risk or failure rate) is obtained by

$$h = 1 - \{1 - p_x(b)\}/\{1 - p_{x-1}(b)\} \tag{10}$$

Replacement model. In this model, a 1-crack state exists in which the distribution of crack length is equal to the initial distribution $\mathbf{P}_0$ in the very portion after inspection and replacement. The probability vector of the crack state after inspection and replacement (at the time $x_1 = 01$) is indicated by

$$\mathbf{P}'_{01} = \{p'_{01}(1), p'_{01}(2), \ldots, p'_{01}(b)\} \tag{11}$$

If the probability of detection and replacement is denoted by $\mathbf{P}_{r1}$, this $\mathbf{P}_{r1}$ is given by the sum of the product of the probability of taking the states $1, 2, \ldots, b$ and the probability of detecting them. Therefore,

$$P_{r1} = \sum_{j=1}^{b} D_j p_x(j) \tag{12}$$

Accordingly, the probability of the existence of the crack length a_j at the time of replacement, $p'_{01}(j)$, is given by the sum of the probability of reproducing the initial crack state $p_0(j)$ by detecting and replacing once and the probability of not detecting in the inspection, in spite of the existence of a crack length a_j at the time x_1. That is,

$$p'_{01}(j) = P_{r1} p_0(j) + (1 - D_j) p_{x1}(j) \tag{13}$$

After inspection and replacement, the probability vector of the crack state at the time x is

$$\mathbf{P}'_x = \{p'_x(1), p'_x(2), \ldots, p'_x(b)\} = \mathbf{P}'_{01}\mathbf{P}^{x-x1} \tag{14}$$

and the probability of fracture is

$$P_f = p'_x(b) \tag{15}$$

The Case Accompanied by a Pre-service Inspection (PSI)

In this case, the probability of detecting defects is

$$D'_j = \text{Prob[detecting crack} \mid \text{crack length } a_j] \tag{16}$$

$$\begin{aligned} D'_j &= 0 \quad (j = 1, 2, \ldots, i-1) \\ 0 \le D'_j &\le 1 \quad (j = i, i+1, \ldots, b) \end{aligned} \tag{17}$$

Repair model. The probability vector of the crack state after PSI is

$$\mathbf{P}_{00} = \{p_{00}(1), p_{00}(2), \ldots, p_{00}(b)\} \tag{18}$$

(The first 0 in the subscript 00 represents the initial distribution and the second 0 indicates pre-service.) Here, in a similar manner to that in Eqn. (7), it follows that

$$P_{00}(j) = (1 - D_j')p_0(j) \tag{19}$$

The probability vector of the crack state after inspection at the time x_1 is

$$\mathbf{P}_{x1} = \mathbf{P}_{00}\mathbf{P}^{x1} \tag{20}$$

The subsequent procedure is the same as that for the ISI case.

Replacement model. The probability vector of the crack state after PSI is

$$\mathbf{P}_{00}' = \{p_{00}'(1), p_{00}'(2), \ldots, p_{00}'(b)\} \tag{21}$$

and, in a similar manner to that in Eqn. (13),

$$p_{00}'(j) = P_{r0}p_0(j) + (1 - D_j')p_0(j) \tag{22}$$

where

$$P_{r0} = \sum_{j=1}^{b-1} D_j' p_0(j) \tag{23}$$

The probability vector of the crack state after inspection at the time x_1 is

$$\mathbf{P}_{x1} = \mathbf{P}_{00}'\mathbf{P}^{x1} \tag{24}$$

(The subsequent procedure is the same as that for the ISI case.)

The procedure for the computation just described is indicated in the flow chart of Fig. 2.

METHOD OF ANALYSIS

Here, it is assumed that the thickness of the plate is 20 mm, that is, the critical crack length corresponding to the critical crack state b is 20 mm. The data [12] indicated in Fig. 3 are used as the initial distribution $\mathbf{P}_0$ of the crack state, this being tabulated in Table 1. The

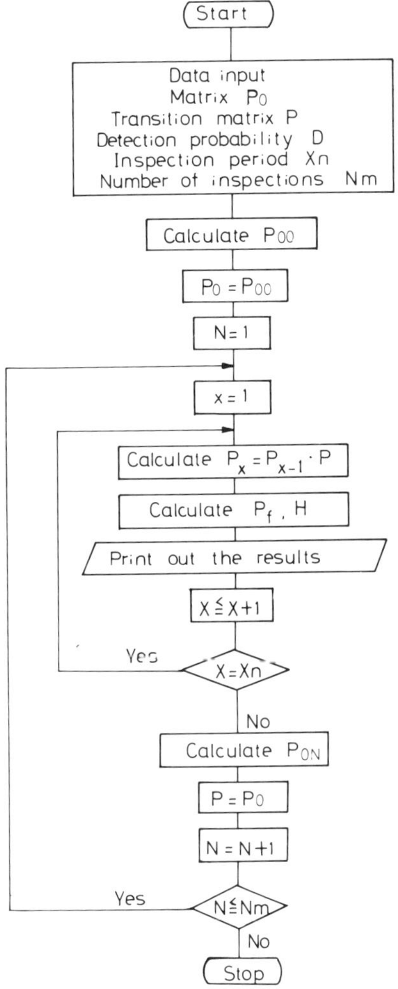

FIG. 2. Flow chart for computation.

initial crack length a_0 is assumed to be 5 mm, corresponding to the initial crack state 1. As we had no adequate datum for crack propagation, the so-called Paris–Erdogan equation

$$\frac{\mathrm{d}a}{\mathrm{d}N} = C(\Delta K)^m$$

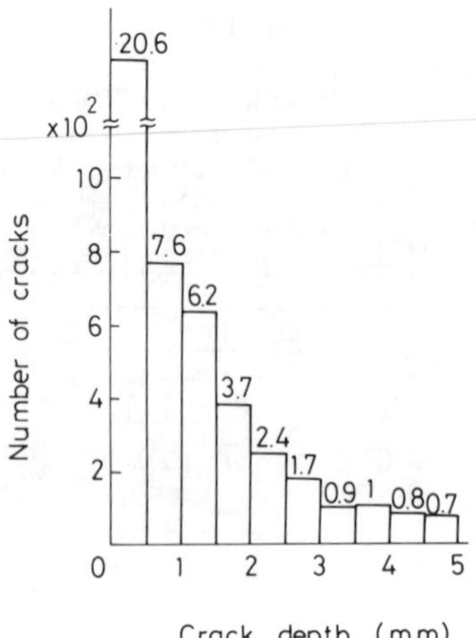

FIG. 3. Data for cracks (after Kashiwagi [12]).

was applied, and the data were created by introducing the values $C = 1{\cdot}062 \times 10^{-13}$ and $\Delta\sigma = 156{\cdot}8$ MPa (stress range). The mean value of the number of stress cycles N corresponding to the crack length $a(N)$ is indicated in Table 2, where, in the case of the analysis using the stochastic transition matrix, if the number of stress cycles is directly used as the number of steps x, the time of computation would be too long. Consequently, it is taken that 2000 cycles is equal to 1

TABLE 1
Assumed distribution of initial cracks

a (mm)	*Probability*
0·5	0·452
1·0	0·167
1·5	0·136
2·0	0·081
2·5	0·053
3·0	0·037
3·5	0·020
4·0	0·022
4·5	0·017
5·0	0·015

TABLE 2
Mean value of the number of cycles to the specified crack length

a (mm)	*N (cycles)* $\times 10^5$
1·0	3·362
1·5	5·241
2·0	6·200
2·5	6·855
3·0	7·388
3·5	7·714
4·0	8·017
4·5	8·267
5·0	8·480
20·0	10·440

step; that is, $N = 2000x$. As in the previous paper [5], the results were obtained such that the process for fatigue crack propagation is the process in which $r_j = p_j/q_j = \sigma_x^2/\bar{x} = \text{const.}$, where σ_x is the standard deviation and $\bar{x}$ is the mean of x. Here, the computation was carried out by taking $r = 5$ and 10. The computed values of $\bar{x}$ and σ_x of the number of steps x are listed in Tables 3 and 4. In the previous paper, it was obtained that, for the fatigue life N_f corresponding to the crack state d,

TABLE 3
Mean, variance and coefficient of variation of the crack propagation life ($r = 5$)

a (mm)	$\bar{x}$	σ_x^2	$\sigma_x/\bar{x}$
0·5	0·0	0	—
1·0	181·6	908	0·166
1·5	262·1	1311	0·138
2·0	310·0	1550	0·127
2·5	342·8	1714	0·121
3·0	366·9	1835	0·117
3·5	385·7	1929	0·114
4·0	400·9	2004	0·112
4·5	413·4	2067	0·110
5·0	424·0	2120	0·109
20·0	522·0	2610	0·098

TABLE 4
Mean, variance and coefficient of variation of the crack propagation life ($r = 10$)

a (mm)	$\bar{x}$	σ_x^2	$\sigma_x/\bar{x}$
0·5	0·0	0	—
1·0	181·6	1816	0·235
1·5	262·1	2621	0·195
2·0	310·0	3100	0·180
2·5	342·8	3428	1·171
3·0	366·9	3669	0·165
3·5	385·7	3857	0·161
4·0	400·9	4009	0·158
4·5	413·4	4134	0·156
5·0	424·0	4240	0·154
20·0	522·0	5220	0·138

expected value:

$$E(N_f) = d - 1 + \sum_{j=1}^{d-1} r_j \tag{25}$$

variance:

$$\mathrm{Var}(N_f) = \sum_{j=1}^{d-1} (1 + r_j)r_j \tag{26}$$

If we calculate the integer value d satisfying the mean value of 522·0 for the number of steps when reaching the critical crack length $a = 20$ mm by using Eqn. (25) in the case of $r = 5$ and 10, the values of d are equal to 88 and 49, respectively. These values give the size of the matrix. From Eqns. (25) and (26), the values of r_j and p_j for each value of j were computed and are listed in Tables 5 and 6. Concerning each element in the vector of the initial distribution of the crack state, the probability of taking the value of a crack length corresponding to the crack states $1, 2, \ldots, b$ can be computed from Table 1.

In the present paper, the probabilities of detecting defects D were assumed to be 0·9 and 0·99, and the critical crack length capable of detection was taken as 2, 3, 4 and 5 mm. The number of inspections was taken as (the number of PSI) + (the number of ISI) between 0 and 6×10^5 cycles.

TABLE 5
Elements in the stochastic matrix ($r = 5$)

j	$r_j = p_j/q_j$	p_j
1–30	5·053	0·8348
31–44	4·750	0·8261
45–52	4·988	0·8330
53–57	5·560	0·8476
58–61	5·025	0·8340
62–64	5·267	0·8404
65–67	4·067	0·8026
68, 69	5·250	0·8400
70, 71	4·300	0·8113
72–87	5·125	0·8367
88	—	1·0000

TABLE 6
Elements in the stochastic matrix ($r = 10$)

j	$r_j = p_j/q_j$	p_j
1–16	10·350	0·9119
17–24	9·063	0·9006
25–29	8·580	0·8956
30–32	9·933	0·9085
33, 34	11·050	0·9170
35, 36	8·400	0·8936
37	14·200	0·9342
38	11·509	0·9200
39	9·600	0·9057
40–48	9·889	0·9082
49	—	1·0000

RESULTS AND DISCUSSION

Figure 4 shows the relationship between the probability of fracture P_f and the number of stress cycles N in the case of $D = 0{\cdot}9$ and $a_d = 3$ mm for the repair model, the parameter being the number of inspections. From Tables 3 and 4, it can be seen that the probability of fracture is higher in the case of $r = 10$ than in the case of $r = 5$ because

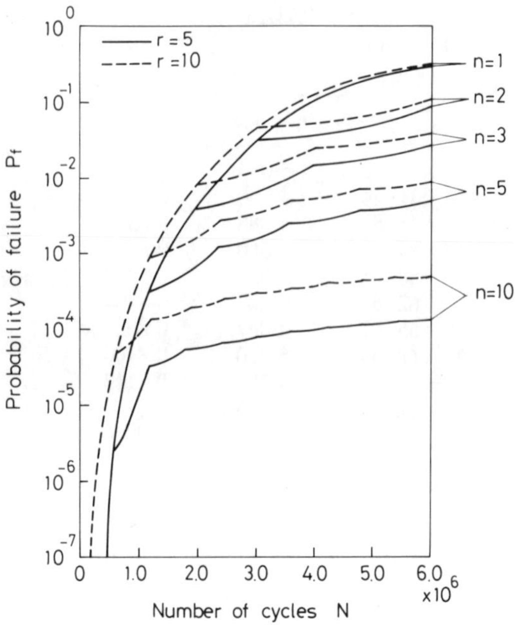

FIG. 4. P_f–N curves (comparison between $r = 5$ and 10 with the repair model).

of the larger variance. The results of computation indicate that the value for the probability of fracture is reduced and the difference between its value for $r = 5$ and $r = 10$ increases with increasing number of inspections.

Figures 5 and 6 show the relationship between the hazard rate h and the number of stress cycles N (the parameter is n) for $r = 5$ in Fig. 5 and $r = 10$ in Fig. 6. The hazard rate decreases with the increasing number of inspections and, while both curves are similar in form, they are different in height.

Figure 7 also indicates the P_f–N relationship and compares the repair model (solid line) with the replacement model (broken line) in the case of $D = 0{\cdot}9$, $a_d = 3$ mm and $r = 5$. From this figure, it can be seen that the probability of fracture in the repair model is lower than that in the replacement model, and this tendency becomes more pronounced as the number of inspections increases. Figure 8 also shows the P_f–N relationship in which $D = 0{\cdot}9$ and $a_d = 5$ mm and $r = 5$, and compares the initial crack length $a_0 = 3$ mm (solid line) with

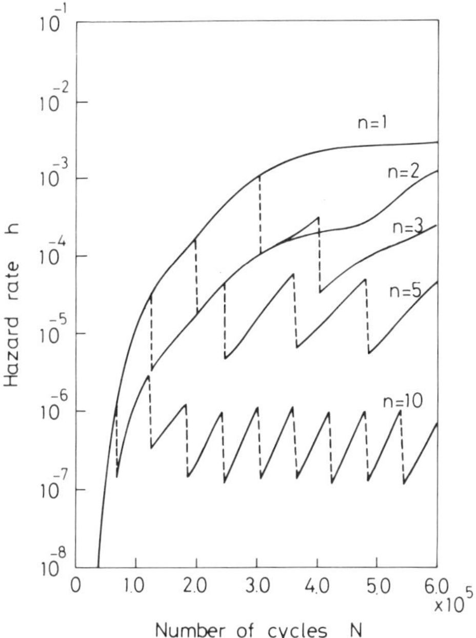

FIG. 5. Hazard rate versus number of cycles ($r = 5$).

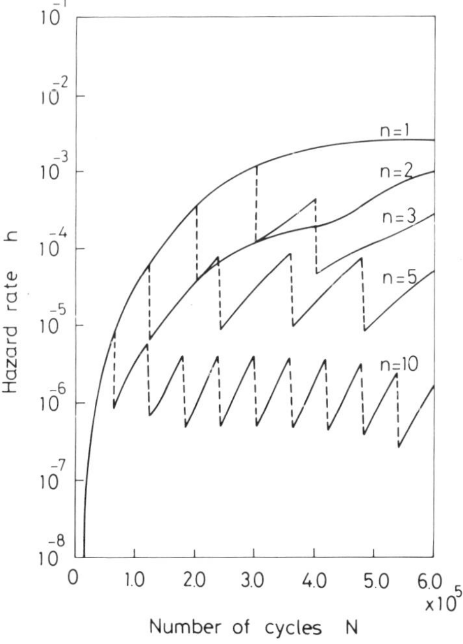

FIG. 6. Hazard rate versus number of cycles ($r = 10$).

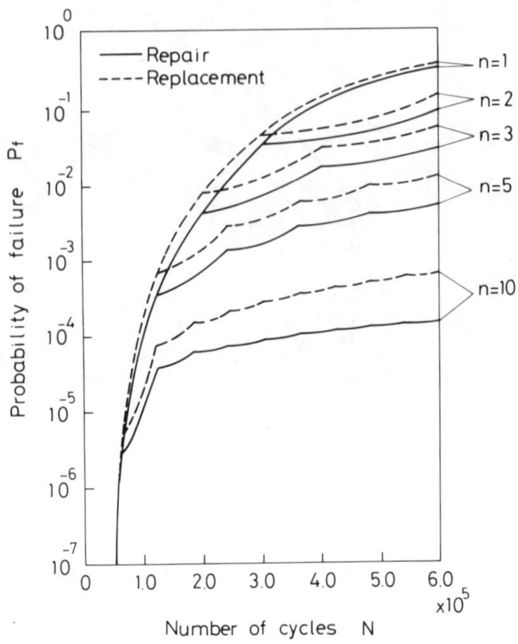

FIG. 7. P_f–N curves (comparison between the repair and replacement models).

$a_0 = 4$ mm (broken line). A remarkable difference can be recognized between both probabilities with a small number of stress cycles, but this difference is reduced as the number of stress cycles increases.

The P_f–N relationship (the parameter is n) shown in Fig. 9 is for the case of $r = 5$, comparing $D = 0{\cdot}9$ and $a_d = 3$ mm (solid line) with $D = 0{\cdot}99$ and $a_d = 5$ mm (broken line). The curves approach each other as the number of inspections increases, and the probability of fracture is very close at $n = 10$. This reveals an aspect aimed at with NDI from the viewpoint of cost.

CONCLUSION

The application of our previous work involving Bogdanoff's theory has been presented, the concept of the Markov chain being applied to the process of fatigue crack propagation. The analysis was performed by taking the ratio of the elements in the stochastic transition matrix, $r = p/q$ to be constant. The application of this theory has been

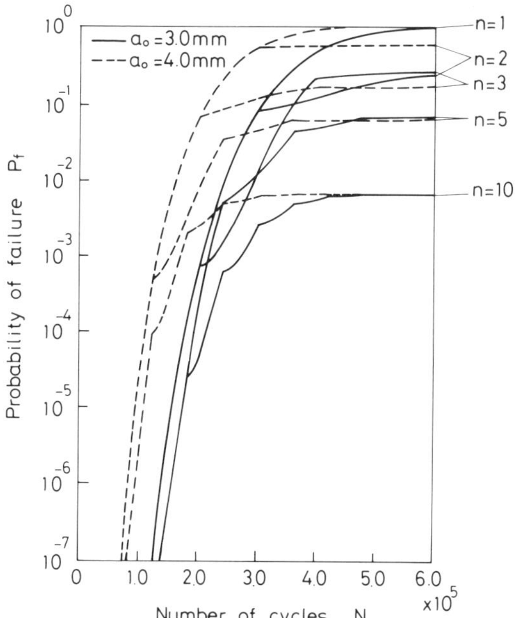

FIG. 8. P_f–N curves (comparison between the initial crack lengths $a_0 = 3$ and 4 mm).

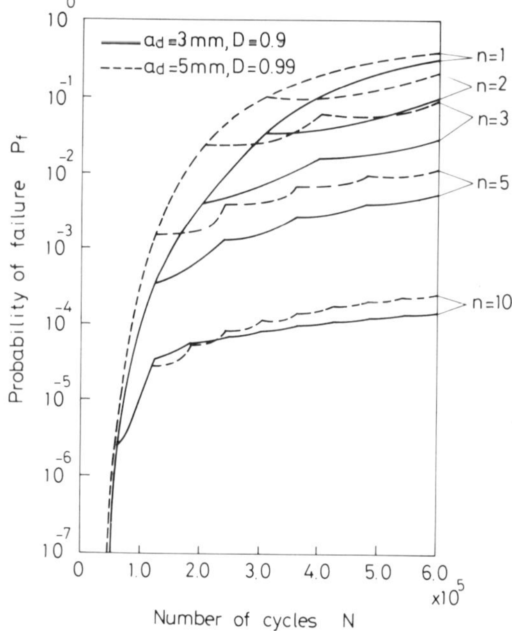

FIG. 9. P_f–N curves (comparison between $a_d = 3$ mm and $D = 0{\cdot}9$, and $a_d = 5$ mm and $D = 0{\cdot}99$).

described for the cases of combining non-destructive inspection (NDI) with ISI and with PSI.

Some analytical examples and the results of computation were shown, and a number of conclusions can be drawn.

(1) The technique for reliability analysis by the Markov chain for fatigue crack propagation proved applicable to the case accompanied by NDI.
(2) The difference in the probability of fracture between $r = 5$ and 10 in the repair model increased with the increasing number of inspections, the probability of fracture being lower in the former case than in the latter.
(3) The hazard rate trend was similar between $r = 5$ and 10, but the absolute values were somewhat different.
(4) The probability of fracture in the repair model was lower than that in the replacement model, the difference between both models increasing as the number of inspections was increased.
(5) The effect of the initial crack length created a higher probability of fracture with increasing initial length. The difference was noticeable at the early stage of fatigue life, and increased with the increase in number of stress cycles.
(6) In the cases of the minimum detectable crack length and the low probability of detecting cracks, the minimum crack length was large and the probability of fracture was low when compared to the case of a high probability for detecting cracks, but there was no difference between these cases as the number of inspections increased.

REFERENCES

[1] J. C. Halpin, T. A. Johnson and M. E. Waddoups, AFML, TR-72-289 W-D AFB (1972).
[2] D. G. Ford, ARL-STRUC., REP-369 (1978); REP-379 (1979); REP-382 (1982).
[3] J. L. Bogdanoff, *J. Appl. Mech.*, **45,** 246 (1978); **45,** 733 (1978). J. L. Bogdanoff and W. Krieger, *J. Appl. Mech.*, **45,** 251 (1978). J. L. Bogdanoff and F. Kozin, *J. Appl. Mech.*, **47,** 40 (1980).
[4] J. L. Bogdanoff and F. Kozin, *Eng. Fract. Mech.*, **14,** 59 (1981).
[5] Y. Shimada, T. Nakagawa and H. Tokuno, *J. Soc. Mater. Sci. Jap.*, **33,** 475 (1984).

[6] Y. Shimada, T. Nakagawa and H. Tokuno, *Preprint, 5th Symp. on Reliability Engineering in Design,* 6 (1983).
[7] H. Itagaki, F. Ozaki and T. Nemoto, *J. Soc. Naval Arch. Jap.,* **139,** 307 (1976).
[8] H. Itagaki, M. Toyoda, A. Mazima and H. Asada, *J. Soc. Naval Arch. Jap.,* **144,** 297 (1981).
[9] H. Kitagawa and T. Hisada, Preprint, Jap. Soc. Mech. Engrs., No. 760-12, 320 (1976); No. 770-2, 300 (1979); No. 770-11, 65 (1977).
[10] H. Kitagawa and T. Hisada, *Trans. Jap. Soc. Mech. Engrs.,* **45,** 1033 (1979).
[11] For example, H. Shiomi, *Introduction to Reliability Engineering,* Maruzen Publishers Ltd., Tokyo, 96 (1967). (in Japanese).
[12] Y. Kashiwagi, *J. Jap. Soc. Mech. Engrs.,* **80,** 545 (1977).

Reliability Analysis of Fatigue Crack Growth Processes under Stationary Random Loading

AKIRA TSURUI
Department of Applied Mathematics and Physics, Kyoto University, Kyoto, Japan

and

AKIRA SAKO
Mitsubishi Heavy Industry Ltd., Hiroshima Works, Hiroshima, Japan

ABSTRACT

To evaluate the effect of in-service inspections on structural reliability, the fatigue crack growth processes under stationary random loading were analyzed on the basis of the Paris–Erdogan crack growth law. A brief survey was first made of the procedures to obtain the crack length distribution with the aid of an extended Markov approximation method, and the result was then applied to investigate how the reliability of structures behaves under periodic inspections.

INTRODUCTION

Since the damage-tolerant design philosophy was widely introduced for such important structures as aircraft, nuclear power plants and ocean structures, the importance of studying the fatigue crack growth processes has undoubtedly increased. Especially, a good understanding of its statistical aspects will be essential for reliability-based design or the maintenance of structures [1]. Further, in order to maintain a high reliability level, it might be necessary to detect cracks by inspection before their size reaches a certain value [2–4].

In this study, the procedures to obtain the crack length distribution under stationary random loading have been briefly surveyed, and by using a Markov approximation method that was developed recently [5, 6], a generalized Fokker–Planck equation will be derived and solved in a closed form. The effect of nondestructive in-service inspection on the reliability degradation due to fatigue crack growth will then be evaluated based upon the result obtained, and comments will be made on how the reliability degrades under periodic inspections.

CRACK LENGTH DISTRIBUTION

We will briefly survey the procedures to derive the crack length distribution by using an extended Markov approximation method [5, 6].

Consider the stochastic crack growth model

$$\frac{\mathrm{d}X}{\mathrm{d}n} = \varepsilon F(S_n, X) \tag{1}$$

where X is the nondimensional crack length, S_n the random stress amplitude after n cycles, ε a small positive parameter and F a given function introduced empirically. Generally speaking, the random stress amplitudes S_n and $S_{n+n'}$ will be correlated with each other. This correlation, however, usually decays as the difference n' tends to increase. Here, we define the correlation time as the difference n_c beyond which the correlation is negligibly small. When we are only interested in the crack length increment after a sufficiently long time compared to the correlation time n_c, the standard practice is to utilize a Markov approximation method.

However, in this problem, careful attention must be paid to the fact that there might be a case in which the crack length tends to infinity after a limited number of stress reversals. In order to overcome this difficulty, the death point $\{\infty\}$ needs to be involved in the state space, this point corresponding to the state at which the cracks, once developed, remain there forever.

Let $X(n)$ be the nondimensional crack length after n cycles of stressing.

$$\Pr[X(n) \in \{\infty\}] = P_\infty(n) \tag{2}$$

represents the probability that $X(n)$ lies at the death point $\{\infty\}$, and

$$\Pr[X(n) \le x \mid X(0) = x_i] = \mathrm{W}(x, n \mid x_i) \tag{3}$$

is the probability that a crack of initial size x_i will grow to a finite value less than or equal to x after n cycles of stressing. Then it is evident that the following relationship holds:

$$\lim_{x \to \infty} W(x, n \mid x_i) + P_\infty(n) = 1 \tag{4}$$

Suppose that the function $W(x, n \mid x_i)$ is differentiable with respect to x to have the derivative $w(x, n \mid x_i)$. For a number of cycles n sufficiently large in comparison with the correlation time n_c, a partial differential equation to describe the variation of $w(x, n \mid x_i)$ can be derived in the following form by the use of a perturbation method, details of which have been discussed in another paper [5].

$$\frac{\partial w}{\partial n} = -\varepsilon \frac{\partial}{\partial x}\{M(x)w\} + \frac{\varepsilon}{2}\frac{\partial^2}{\partial x^2}\{D(x)w\} \tag{5}$$

$$\left.\begin{aligned} M(x) &= E^*[F(S_n, x)] + \varepsilon \int_{-\infty}^{0} \left\{E^*\left[\frac{\partial F(S_n, x)}{\partial x} F(S_{n+n'}, x)\right]\right. \\ &\quad \left. - E^*\left[\frac{\partial F(S_n, x)}{\partial x}\right] E^*[F(S_{n+n'}, x)]\right\} \mathrm{d}n' \\ D(x) &= 2\varepsilon \int_{-\infty}^{0} \{E^*[F(S_n, x)F(S_{n+n'}, x)] \\ &\quad - E^*[F(S_n, x)]E^*[F(S_{n+n'}, x)]\} \,\mathrm{d}n' \end{aligned}\right\} \tag{6}$$

where the notation $E^*[f(Y)]$ is defined by

$$E^*[f(Y)] = \frac{E[f(Y)\exp(-\alpha Y)]}{E[\exp(-\alpha Y)]} \tag{7}$$

in which α is an arbitrary positive constant satisfying $\alpha = O(\varepsilon^3)$. Equation (5) can be regarded as the Fokker–Planck equation in a generalized sense.

With the aid of the Paris–Erdogan crack growth model, the most famous at present, Eqn. (1) reduces to

$$\frac{\mathrm{d}X}{\mathrm{d}n} = \varepsilon Z_n^{\nu} X^{\nu/2} \tag{8}$$

where $Z_n = S_n/s_0$, and ν stands for a material constant usually assuming a value not less than 2·0. Therefore, $M(x)$ and $D(x)$ defined by Eqn. (6) can be given more specifically as

$$\left.\begin{aligned} M(x) &= E[Z_n^\nu]x^{\nu/2} + \frac{\varepsilon\nu}{2}\int_{-\infty}^{0} \{E[Z_n^\nu Z_{n+n'}^\nu] \\ &\quad - E[Z_n^\nu]E[Z_{n+n'}^\nu]\}\,\mathrm{d}n' x^{\nu-1} \\ D(x) &= 2\varepsilon\int_{-\infty}^{0} \{E[Z_n^\nu Z_{n+n'}^\nu] - E[Z_n^\nu]E[Z_{n+n'}^\nu]\}\,\mathrm{d}n' x^\nu \end{aligned}\right\} \tag{9}$$

Introducing a new variable

$$\tau = \varepsilon E[Z_n^\nu]n \tag{10}$$

and a new parameter

$$\mu = \frac{2\varepsilon}{E[Z_n^{2\nu}]}\int_{-\infty}^{0} \{E[Z_n^\nu Z_{n+n'}^\nu] - E[Z_n^\nu]E[Z_{n+n'}^\nu]\}\,\mathrm{d}n' \tag{11}$$

the Fokker–Planck equation (5) reduces to the following simple form:

$$\frac{\partial w}{\partial \tau} = -\frac{\partial}{\partial x}\left\{\left(x^{\nu/2} + \frac{\mu\nu}{4}x^{\nu-1}\right)w\right\} + \frac{\mu}{2}\frac{\partial^2}{\partial x^2}\{x^\nu w\} \tag{12}$$

Further, for later convenience, the correlation time n_c is defined by

$$n_c = \frac{2\int_{-\infty}^{0} \{E[Z_n^\nu Z_{n+n'}^\nu] - E[Z_n^\nu]E[Z_{n+n'}^\nu]\}\,\mathrm{d}n'}{E[Z_n^{2\nu}] - [E[Z_n^\nu])^2} \tag{13}$$

The emphasis should be placed here on the case in which the following relationship holds:

$$\frac{\mu}{\tau} = \frac{n_c}{n}\left\{\frac{E[Z_n^{2\nu}]}{(E[Z_n^\nu])^2} - 1\right\} \ll 1 \tag{14}$$

After some mathematical manipulation, the solution which satisfies

$$\lim_{\tau\to 0} w(x, \tau \mid x_i) = \delta(x - x_i) \tag{15}$$

of the simplified Fokker–Planck equation (12) can be given as

$$w(x, \tau \mid x_i) = \frac{1}{x^{\lambda+1}\sqrt{2\pi\mu\tau}}\exp\left\{-\frac{(x_i^{-\lambda} - x^{-\lambda} - \lambda\tau)^2}{2\lambda^2\mu\tau}\right\} \tag{16}$$

where

$$\lambda = \frac{\nu}{2} - 1 \tag{17}$$

and δ is a Dirac delta function.

In the limit of $\lambda \to 0$, Eqn. (16) reduces to

$$w(x, \tau \mid x_i) = \frac{1}{x\sqrt{2\pi\mu\tau}} \exp\left[-\frac{\{\ln(x\mathrm{e}^{-\tau}/x_i)\}^2}{2\mu\tau} \right] \tag{18}$$

which is a log-normal distribution. It finally follows from Eqn. (16) that, for a general case in which

$$\lim_{\tau \to 0} w(x, \tau) = g(x) \tag{19}$$

is satisfied, the solution of the simplified Fokker–Planck equation (12) can be written as

$$w(x, \tau) = \frac{1}{x^{\lambda+1}\sqrt{2\pi\mu\tau}} \int_0^\infty \exp\left\{ -\frac{(x_i^{-\lambda} - x^{-\lambda} - \lambda\tau)^2}{2\lambda^2\mu\tau} \right\} g(x_i)\, \mathrm{d}x_i \tag{20}$$

INSPECTION EFFECT

Making use of the results obtained in the preceding section, we will now describe the behavior of fatigue crack growth under a non-destructive in-service inspection program, and discuss the effect of the inspection on the structural reliability of the component. Let T be a random variable which expresses the time for the crack size to reach a prescribed critical value x_c for the first time, and let $H(\tau)$ be its distribution function; i.e.

$$H(\tau) = \Pr[T \le \tau] \tag{21}$$

Using Eqn. (20), we can derive the following approximate relationship [5, 6] in the case of no inspections:

$$H(\tau) = H_0(\tau) \equiv 1 - \int_0^{x_c} w(x, \tau)\, \mathrm{d}x \tag{22}$$

where x_c is the maximum crack length beyond which the structural component will fail. In what follows, we assume that the initial crack length must have been less than x_c, otherwise the component would have failed initially. Considering structures which are expected to have a very high level of reliability, we usually require the failure probability $H(\tau)$ up to time τ to be very small, while it rapidly increases as the time elapses. Thus, in order to keep the structures highly reliable, it is standard practice to inspect the structural components and to exchange them if cracks are detected in the inspections.

Let $D(x)$ be the crack detection probability for a crack size x, and τ^* be the time of inspection. From Eqn. (20), we have a density function $w^*(x)$ of crack size x at the inspection time τ^* as follows:

$$w^*(x) = \begin{cases} \dfrac{w(x, \tau^*)}{1 - H(\tau^*)} & (x \leq x_c) \\ 0 & (x_c < x) \end{cases} \tag{23}$$

The crack detection probability D of the inspection time τ^* follows from this equation as

$$D = \int_0^{x_c} D(x) w^*(x)\, dx \tag{24}$$

If no cracks are detected in this inspection, the density function of the crack length yields

$$v(x) = k\{1 - D(x)\} w^*(x) \tag{25}$$

where k stands for the normalizing constant that makes the integral of $v(x)$ in the range $(0, x_c)$ unity.

In a case in which the component is exchanged, assume that the density function of a crack length contained in the new component also equals $g(x)$. Therefore, considering that the process $X(\tau)$ is of a Markov type, we can deduce the conditional residual life distribution $H_{XC}(\tau)$, given that the component has been exchanged at time τ^*, if the variable τ in Eqn. (22) is replaced by $\tau - \tau^*$, because the state immediately after the exchange is the same as the state at $\tau = 0$.

$$H_{XC}(\tau) = 1 - \int_0^{x_c} \int_0^{x_c} w(x, \tau - \tau^* \mid x_i) g(x_i)\, dx_i\, dx \quad (\text{for } \tau > \tau^*) \tag{26}$$

On the other hand, in a case in which no cracks are detected, and replacing $g(x)$ in Eqn. (22) by $v(x)$, we have the conditional residual life distribution $H_{UD}(\tau)$, and assuming that no cracks could be found in the inspection,

$$H_{\mathrm{UD}}(\tau) = 1 - \int_0^{x_c} \int_0^{x_c} w(x, \tau - \tau^* \mid x_i) v(x_i) \, dx_i \, dx \quad (\text{for } \tau > \tau^*) \tag{27}$$

Equations (26) and (27) lead to the residual life distribution $H(\tau)$ for $\tau > 0$ of

$$H(\tau) = \begin{cases} H_0(\tau) & (\tau \le \tau^*) \\ H_0(\tau^*) + \{1 - H_0(\tau^*)\}\{D H_{\mathrm{XC}}(\tau) + (1 - D) H_{\mathrm{UD}}(\tau)\} & (\tau > \tau^*) \end{cases} \tag{28}$$

This result, however, would be too complicated to be easily interpreted, unless the function $H(\tau)$ could be numerically calculated.

Now, we will fix the functions $D(x)$ and $g(x)$. Assume that the time to crack initiation obeys a two-parameter Weibull distribution $W(t_0)$ [2].

$$W(t_0) = 1 - \exp\left\{ -\left(\frac{t_0}{\beta}\right)^{\gamma} \right\} \tag{29}$$

Then, the density function $g(x)$ of the initial crack size (Fig. 1) may be derived from this equation through

$$\begin{aligned} g(x) \, dx &= -dW(t_0) \\ t_0 &= c(\tilde{x}^{-\lambda} - x^{-\lambda}) \end{aligned} \tag{30}$$

where c and $\tilde{x}$ denote suitable constants.

There are several methods for detecting cracks, including, X-ray, dye-penetrant, ultrasonic and magnetic. Among these, we will consider the ultrasonic inspection method in this study, which can be expected to be highly sensitive. Harris and Lim [7] proposed the law

$$D(x) = 1 - \tfrac{1}{2} \operatorname{erfc}\left\{ K \ln\left(\frac{x}{x^*}\right) \right\} \tag{31}$$

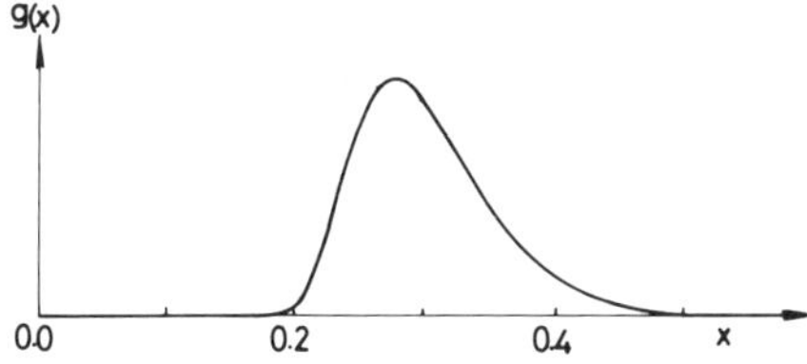

FIG. 1. Probability density function $g(x)$ of the initial crack length.

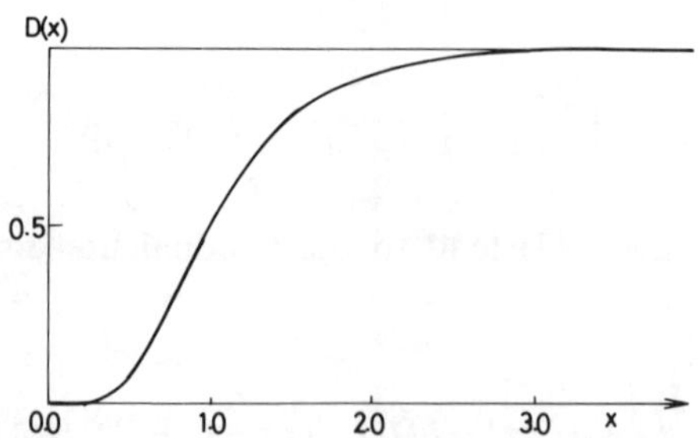

FIG. 2. Crack detection probability as a function of the crack size ($K = 2{\cdot}1$, $x^* = 1{\cdot}0$).

for ultrasonic inspection (Fig. 2), where K and x^* are suitable constants.

The behavior of the residual life distribution $H(\tau)$ is shown in Fig. 3, in which the ordinate is of logarithmic scale. The solid and broken lines correspond to $H(\tau)$ with and without inspection, respectively, and small circles represent the inspection time. We can see from this figure that the effect of inspection is temporary and gradually decreases as time elapses. Hence, we cannot expect a significant effect unless the inspection is repeated.

RELIABILITY DEGRADATION UNDER PERIODIC INSPECTIONS

In the preceding section, we saw that a significant effect from inspection could not be expected unless the inspection was repeated.

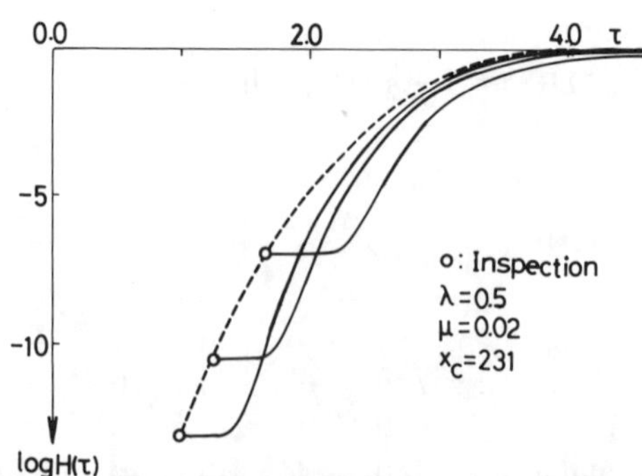

FIG. 3. Reliability degradation after inspection.

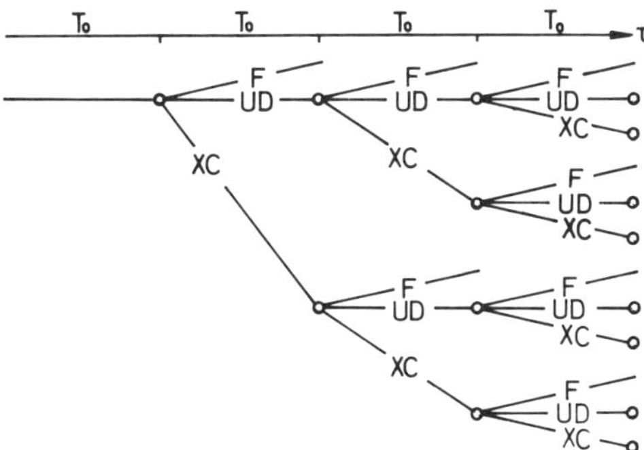

FIG. 4. Schematic representation of state transitions with time.

So, in this section, we will discuss the reliability degradation under periodic inspections.

Suppose that the interval between successive inspections is T_0, whether the component is exchanged or not. This aspect is shown schematically in Fig. 4, in which UD means that no crack could be detected by the inspection, XC means that at least one crack was detected and the component was exchanged and F means that the component has failed in the interval T_0.

Taking the approximate Markov property of the crack growth process and the regularity of the inspection interval into account, we can describe this process as a Markov chain, if our main concern is for the state at the inspection time only. The states are defined as follows: a_0 is the state immediately after a component has been exchanged, a_i $(i = 1, 2, \ldots)$ are the states in which no cracks were detected by i-times successive inspections since the component was exchanged and f is the state in which the component has failed. Figure 5 shows schematically how transitions between the states take place.

Then, we may consider a Markov chain specified in terms of the state probability vector

$$\mathbf{b}_n = (b_{\mathrm{f}}^{(n)}, b_0^{(n)}, b_1^{(n)}, \ldots, b_i^{(n)}, \ldots) \tag{32}$$

and the Markov matrix

$$Q = \begin{pmatrix} 1 & 0 & 0 & 0 & 0 & 0 & \cdots \\ p_{0\mathrm{f}} & p_{00} & p_{01} & 0 & 0 & 0 & \cdots \\ p_{1\mathrm{f}} & p_{10} & 0 & p_{12} & 0 & 0 & \cdots \\ p_{2\mathrm{f}} & p_{20} & 0 & 0 & p_{23} & 0 & \cdots \\ p_{3\mathrm{f}} & p_{30} & 0 & 0 & 0 & p_{34} & \cdots \\ \vdots & \vdots & \vdots & \vdots & \vdots & \vdots & \cdots \end{pmatrix} \tag{33}$$

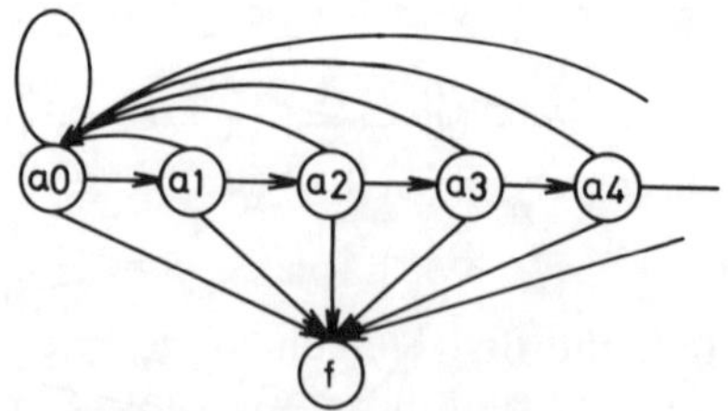

FIG. 5. Diagram representing the transition between states.

where $b_f^{(n)} = \Pr$[the component is in the state f at time nT_0] and $b_i^{(n)} = \Pr$[the component is in the state a_i at time nT_0] for $i = 0, 1, 2, \ldots$.

Introducing a series of functions $v_i(x)$ through

$$v_i(x) = \frac{\{1 - D(x)\} \int_0^{x_c} w(x, T_0 \mid x_0) v_{i-1}(x_0) \, dx_0}{\int_0^{x_c} \int_0^{x_c} \{1 - D(x)\} w(x, T_0 \mid x_0) v_{i-1}(x_0) \, dx_0 \, dx} \quad (i = 1, 2, \ldots) \tag{34}$$

and

$$v_0(x) = g(x) \tag{35}$$

we can calculate the transition probabilities in Eqn. (33) as follows:

$$p_{ij} = \begin{cases} \int_0^{x_c} \int_0^{x_c} \{1 - D(x)\} w(x, T_0 \mid x_0) \, v_i(x_0) \, dx_0 \, dx & (j = i + 1) \\ \int_0^{x_c} \int_0^{x_c} D(x) w(x, T_0 \mid x_0) v_i(x_0) \, dx_0 \, dx & (j = 0) \\ 0 & (\text{otherwise}) \end{cases} \tag{36}$$

$$p_{if} = 1 - \int_0^{x_c} \int_0^{x_c} w(x, T_0 \mid x_0) v_i(x_0) \, dx_0 \, dx \tag{37}$$

We should note that the function $v_i(x)$ is the probability density function of the crack length just after the ith inspection, given that no crack was detected in the component by successive i-times inspections since it was exchanged.

Consequently, if the initial vector $\mathbf{b}_0$ is given by

$$\mathbf{b}_0 = (0, 1, 0, 0, \ldots) \tag{38}$$

then we have

$$\mathbf{b}_n = \mathbf{b}_0 Q^n \tag{39}$$

It is to be noted that the first element of $\mathbf{b}_n$, i.e. $b_{\mathrm{f}}^{(n)}$, is the failure probability up to time nT_0, and that if we want to know only the values of $b_{\mathrm{f}}^{(n)}$, the functions $v_j(x)$, together with the values

$$\left.\begin{aligned} P_j &= \int_0^{x_c}\int_0^{x_c} D(x)w(x, T_0 \mid x_0)v_{j-1}(x_0)\,\mathrm{d}x_0\,\mathrm{d}x \\ Q_j &= \int_0^{x_c}\int_0^{x_c} \{1 - D(x)\}w(x, T_0 \mid x_0)v_j(x_0)\,\mathrm{d}x_0\,\mathrm{d}x \end{aligned}\right\}(j = 1, 2, \ldots) \tag{40}$$

and a recurrence formula

$$\begin{aligned} w_n(x) &= \sum_{j=1}^{n-1} P_j w_{n-j}(x) + Q_{n-1}\int_0^{x_c} w(x, T_0 \mid x_0)v_{n-1}(x_0)\,\mathrm{d}x_0 \\ &\qquad\qquad\qquad\qquad\qquad\qquad\qquad (n = 2, 3, \ldots) \\ w_1(x) &= \int_0^{x_c} w(x, T_0 \mid x_0)g(x_0)\,\mathrm{d}x_0 \end{aligned} \tag{41}$$

lead us to the simple relationship

$$b_{\mathrm{f}}^{(n)} = 1 - \int_0^{x_c} w_n(x)\,\mathrm{d}x \tag{42}$$

Up to this point, we have referred to the probabilities of the states only at the instant of inspection, which gives sufficient information in most cases. We can, however, easily obtain the residual life distribution $H(\tau)$ for an arbitrary time τ.

$$H(\tau) = \begin{cases} H_0(\tau) & (0 \leq \tau < T_0) \\ 1 - \displaystyle\int_0^{x_c}\int_0^{x_c} w(x, \tau - nT_0 \mid x_0)w_n(x_0)\,\mathrm{d}x_0\,\mathrm{d}x & \\ & (nT_0 \leq \tau < (n+1)T_0) \end{cases} \tag{43}$$

This equation will be used in the numerical examples. Here, we should be aware that the function $w(x, \tau)$ varies rapidly with respect to x when τ is little larger than nT_0, and that this fact may cause a large numerical error.

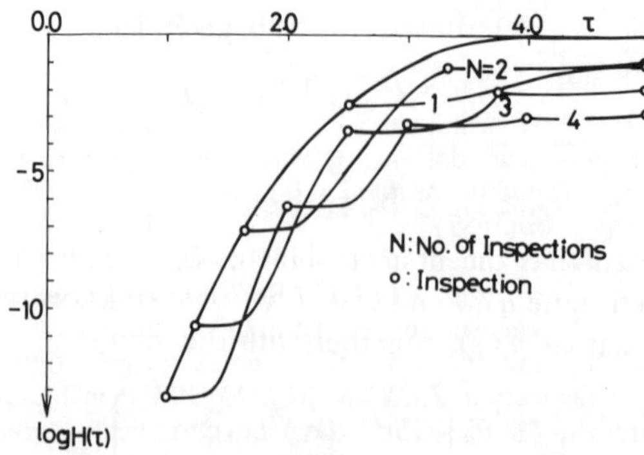

FIG. 6. Reliability degradation between periodic inspections.

Numerical results are plotted in Fig. 6 for $T_0 = 2{\cdot}5$, $1{\cdot}667$, $1{\cdot}25$ and $1{\cdot}0$. In each case, the number of inspections n is 1, 2, 3 and 4, respectively. As might be expected, the larger n becomes the smaller the failure probability which generally results. We can, however, find a curious phenomenon that $H(5{\cdot}0)$ with one inspection is less than $H(5{\cdot}0)$ with two inspections. It might be interpreted as follows: In the case of $n = 1$, when we carry out the first inspection at $\tau = 2{\cdot}5$, the probability that we can find cracks is 92%, so that, in most cases, the component is renewed then. Hence, the probability of failure would be small. On the other hand, in the case of $n = 2$, the crack detection probability at the first inspection is 49%, so that, in a little over half of the cases, the component is left as it is until the second inspection at $\tau = 3{\cdot}333$. Therefore, the probability of failure may be large. Apparently, this is an exceptional case since this phenomenon cannot be seen in the case of $n = 3$ and 4.

CONCLUSIONS

In this study, we have presented an available method to deduce the effect of repeated inspections on the reliability degradation due to fatigue crack growth under random loading based on the Paris–Erdogan law. The method indicates that repeated inspections have a large effect on keeping structures highly reliable.

REFERENCES

[1] See for example: H. Ishikawa, A. Tsurui and H. Kimura, Stochastic fatigue crack growth model and its wide applicability in reliability-based design, *Current Japanese Materials Research,* Vol. 2, Elsevier Applied Science, London, 45 (1987).
[2] M. Shinozuka, Development of reliability-based aircraft safety criteria: an impact analysis, *Tech. Rep. AFFDL-TR-76-36,* Vol. 1, April (1976).
[3] Y. Shimada, T. Nakagawa and H. Tokuno, Application of Markov chain to the reliability analysis of fatigue crack propagation process with nondestructive inspection, *J. JSMS,* **34,** 327 (1985) (in Japanese).
[4] A. Sako and A. Tsurui, The effect of periodical inspections on the reliability degradation due to fatigue crack growth, *Structural Safety and Reliability, Vol. 1* (*Proc. JCOSSAR'87*), The Society of Materials Science, Kyoto, Japan, 153 (1987) (in Japanese).
[5] A. Tsurui and H. Ishikawa, Application of Fokker–Planck equation to a stochastic fatigue crack growth model, *Structural Safety,* **4,** 15 (1986).
[6] A. Tsurui, H. Ishikawa and A. Utsumi, Theoretical study on the distribution of fatigue crack propagation life under stationary random loading, *Proc. FATIGUE 84,* Vol. II, University of Birmingham, 949 (1984).
[7] D. O. Harris and E. Y. Lim, Application of a probabilistic fracture mechanics model to the influence of in-service inspection on structural reliability, *ASTM STP,* **798,** 19 (1983).

Bayesian Reliability Analysis for Evaluating In-Service Inspection

HIROSHI ITAGAKI

Department of Naval Architecture and Ocean Engineering, Yokohama National University, Yokohama, Kanagawa, Japan

SEIICHI ITOH

First Airframe Division, National Aerospace Laboratory, Mitaka, Tokyo, Japan

and

NORIO YAMAMOTO

Development Department, Nippon Kaiji Kyokai, Chiyoda-ku, Tokyo, Japan

ABSTRACT

This report emphasizes the application of the Bayesian method in structural reliability analysis to determine appropriate inspections, which are significant for continuing structural integrity and should be evaluated by a probabilistic approach. However, credible analysis is prevented by the lack of data available *a priori*.

From this viewpoint, reliability analyses have been developed that involve a function for estimating uncertain factors by using the Bayesian method and field data gathered during in-service inspections.

Two numerical examples are described to evaluate the inspection intervals for ship structures and to decide on the proper inspection intervals.

INTRODUCTION

In order to satisfy structural integrity requirements, it is well known that appropriate inspections play an important role. As those inspections are required to detect fatigue cracks prior to propagation to a certain critical crack length, the first inspection and the inspection interval or frequency should be appropriately determined with the aid of reliability analysis, because several probabilistic factors need to be considered such as the initial crack length, fatigue crack initiation and propagation, corrosion, service loads, environment, strength and NDI detectability [1, 2]. However, for a structural reliability analysis to be performed with a high degree of engineering basis, the analytical models used must obviously be defined reasonably well. In detail, variables in those analytical models must be specified in terms of probability density functions (PDFs). Unfortunately, the prevailing paucity of pertinent data prevents the credibility not only of the analytical forms of those PDFs, but also of their parameter values estimated by using such data.

On the other hand, available data derived from in-service inspections involve essential information on the structure inspected by a program, which was determined by the reliability analysis with the uncertain factors mentioned above. It is accepted that the Bayesian method [3] is one of the most effective methods for estimating such factors from data gathered during inspections.

From these viewpoints, the authors have developed several reliability analyses applying the Bayesian method [4–14] and have demonstrated their validity by applying analytical methods, Monte Carlo simulations and field data obtained through in-service inspections.

This report is intended to introduce two examples [13, 14] of the Bayesian method investigated recently by the authors in order to reveal its validity.

APPLICATION OF THE BAYESIAN METHOD TO FIELD DATA

Several reliability analyses have been performed to ensure the continuing integrity of ship structures. However, as significant damage rarely occurs in large-scale structures, it would be difficult to collect effective damage experience. Meanwhile, it is considered that there are many components in a ship's structure which, even if not significant at a comparatively early stage, could be the cause of hazardous failure of the structure. Fatigue cracks emanating at the lower ends of hold frames are a typical example. As a matter of course, the main purpose of inspecting ship structures is to detect those cracks before they are propagated to a critical length, so as to maintain the safety level.

In this example [13], actual inspection data on the lower ends of the hold frames in a ship will be analyzed by the Bayesian method to estimate the safety of the ship's structure after inspection and to evaluate the inspection program performed on these lower ends, which are subsequently referred to as members.

Analytical Model and Formulation

Analytical Model

The following assumptions were made for the model in order to analyze the ship's structure without undue difficulty:

(1) All structural members, which are assumed to belong to the same population, are visually inspected, and when a fatigue crack in a member is detected, it is repaired or replaced to reproduce the initial state.

(2) The fatigue life of a member is defined as the duration between the beginning of service and the time when a fatigue crack is propagated to the minimum length detectable by visual inspection. It is difficult to determine the PDF governing the life *a priori* due to the essential dependence on the structural characteristics. Therefore, in this analysis, an estimation of an uncertain PDF is replaced by estimating an uncertain scale parameter β in the following PDF, which is a two-parameter Weibull distribution.

$$F_T(t \mid \beta) = 1 - \exp\left\{-\left(\frac{t}{\beta}\right)^{\alpha}\right\} \tag{1}$$

(3) The length of a fatigue crack which exists in a member used beyond its fatigue life is assumed to be a random variable, which is highly dependent on the structural characteristics, and is governed by an exponential distribution for simplicity. The parameter λ, which is a function of the service time τ, is considered to be uncertain.

$$F_X(x \mid \lambda) = [1 - \exp\{-\lambda(\tau) \cdot (x - \mu)\}] \qquad x \geq \mu \tag{2}$$

in which μ is the minimum crack length.

(4) The probability of detecting a crack of length x is also assumed

to be an exponential function of x with an uncertain parameter a.

$$D(x \mid a) = 1 - \exp\{-a(x-b)\} \qquad x \geq b \tag{3}$$

where b is the minimum crack length detectable by visual inspection.

Formulation

For the case where cracks have been detected in n_k structural members out of M structural members in a ship at the kth inspection performed at the time t_k, the following events are involved: event A, $E_V[A]$, is a certain length of crack detected, and event B, $E_V[B]$, is no crack detected. $E_V[B]$ involves two events of $E_V[B_1]$ and $E_V[B_2]$. The former is that of an existing crack not detected and the latter is no crack existing.

The length of cracks, detected by visual inspection, in the structural members of a ship is not usually measured. Therefore, in this analysis, an average detectability derived from Eqns. (2) and (3) will be applied.

$$\bar{D}(\tau \mid a, \lambda) = \int_0^\infty D(x \mid a)\{\mathrm{d}F_X(x \mid \lambda)/\mathrm{d}x\}\,\mathrm{d}x \tag{4}$$

Furthermore, $\bar{D}$ is assumed to be an uncertain constant for simplicity.

Probability of event A, $E_V[A]$. Event A means that (1) a member whose service has started at a certain time exceeds its fatigue life in service, and a fatigue crack is initiated in the member; and (2) the crack is detected for the first time at the kth inspection performed at the time t_k. The probability of $E_V[A]$ is given by

$$P\{E_V[A_k \mid \beta, \bar{D}]\} = \sum_{i=0}^{k-1} \frac{n_i}{M} \sum_{j=1}^{k-i} [\{F_T(t_{i+j} - t_i \mid \beta) - F_T(t_{i+j-1} - t_i \mid \beta)\}(1-\bar{D})^{k-(i+j)}\bar{D}] \tag{5}$$

in which $n_0 = M$ and $t_0 = 0$.

Probabilities of events B_1, $E_V[B_1]$, *and* B_2, $E_V[B_2]$. Event B_1 means that (1) a member whose service has started at a certain time exceeds its fatigue life in service, and a fatigue crack is initiated in the member; however, (2) this crack is not detected. The probability of $E_V[B_1]$

becomes

$$P\{E_V[B_{1k} \mid \beta, \bar{D}]\} = \sum_{i=0}^{k-1} \frac{n_i}{M} \sum_{j=1}^{k-i} [\{F_T(t_{i+j} - t_i \mid \beta) - F_T(t_{i+j-1} - t_i \mid \beta)\}(1 - \bar{D})^{k-(i+j-1)}] \tag{6}$$

Event B_2 means that a member whose service has started at a certain time is within its fatigue life at the kth inspection. The probability of $E_V[B_2]$ is given by

$$P\{E_V[B_{2k} \mid \beta]\} = \sum_{i=0}^{k-1} \frac{n_i}{M} \{1 - F_T(t_k - t_i \mid \beta)\} \tag{7}$$

When fatigue cracks are detected in n_k members and not detected in the remaining $(M - n_k)$ at the kth inspection, the probability of this event can be derived from Eqn. (8) with the aid of Eqns. (5)–(7).

$$P[\text{Data}_k \mid \beta, \bar{D}] = (P\{E_V[A_k \mid \beta, \bar{D}]\})^{n_k}(P\{E_V[B_{1k} \mid \beta, \bar{D}]\} + P\{E_V[B_{2k} \mid \beta]\})^{M-n_k} \tag{8}$$

Bayesian Analysis

The scale parameter β in the PDF of the fatigue life and the average detectability $\bar{D}$, which are considered as uncertainties in this analysis, can be estimated by employing the Bayesian method and field data collected during inspections.

An initial prior joint PDF, p^0, of β and $\bar{D}$ is selected to be uniformly distributed over the ranges of β_1 and β_2 and $\bar{D}_1$ and $\bar{D}_2$. It is to be noted that these ranges have been determined on the basis of experience obtained from similar structures.

$$p^0(\beta, \bar{D}) = 1/\{(\beta_2 - \beta_1)(\bar{D}_2 - \bar{D}_1)\} \qquad \beta_2 \geq \beta \geq \beta_1 \qquad \bar{D}_2 \geq \bar{D} \geq \bar{D}_1 \tag{9}$$

When the kth inspection has been completed, one can obtain the following formula for the posterior joint PDF of β and $\bar{D}$ from Eqns. (8) and (9).

$$p^k(\beta, \bar{D}) = P[\text{Data}_k \mid \beta, \bar{D}] \cdot p^0(\beta, \bar{D}) \Big/ \int_{\beta_1}^{\beta_2} \int_{\bar{D}_1}^{\bar{D}_2} [\text{Numerator}] \, d\bar{D} \, d\beta \tag{10}$$

The expected number of cracked members after the time t since completing the kth inspection can now be derived.

First, the expected probability of survival of M members becomes

$$\bar{R}_M(t) = \int_{\beta_1}^{\beta_2} \int_{\bar{D}_1}^{\bar{D}_2} R_M(t \mid \beta, \bar{D}) \cdot p^k(\beta, \bar{D})\, \mathrm{d}\bar{D}\, \mathrm{d}\beta \tag{11}$$

in which, as n_k members with detected fatigue cracks are repaired and the remaining $(M - n_k)$ members are not repaired, $R_M(t \mid \beta, \bar{D})$ can conservatively be given by

$$R_M(t \mid \beta, \bar{D}) = \{R(t;\text{repair} \mid \beta)\}^{n_k}\{R(t;\text{no repair} \mid \beta, \bar{D})\}^{M-n_k} \tag{12}$$

with

$$R(t;\text{repair} \mid \beta) = 1 - F_T(t \mid \beta) \tag{13}$$

$$R(t;\text{no repair} \mid \beta, \bar{D}) = \frac{\sum_{i=1}^{k-1} \frac{n_i}{M}\{1 - F_T(t_k - t_i + t \mid \beta)\}}{P\{\mathrm{E_V}[\mathrm{B}_{1k} \mid \beta, \bar{D}]\} + P\{\mathrm{E_V}[\mathrm{B}_{2k} \mid \beta]\}} \tag{14}$$

Second, using Eqns. (13) and (14), the expected number of cracked members is obtained from

$$E[n'(t);k] = n_k\{1 - \bar{R}_1^k(t)\} + (M - n_k)\{1 - \bar{R}_2^k(t)\} \tag{15}$$

in which

$$\bar{R}_1^k(t) = \int_{\beta_1}^{\beta_2} \int_{\bar{D}_1}^{\bar{D}_2} R(t;\text{repair} \mid \beta) \cdot p^k(\beta, \bar{D})\, \mathrm{d}\bar{D}\, \mathrm{d}\beta \tag{16}$$

$$\bar{R}_2^k(t) = \int_{\beta_1}^{\beta_2} \int_{\bar{D}_1}^{\bar{D}_2} R(t;\text{no repair} \mid \beta, \bar{D}) \cdot p^k(\beta, \bar{D})\, \mathrm{d}\bar{D}\, \mathrm{d}\beta \tag{17}$$

and the expected number of remaining cracked members $E[n';k]$ just after the kth inspection are derived from $t = 0$ in Eqn. (15).

On the other hand, the expected number of cracked members detected becomes

$$E[n^*(t);k] = n_k \cdot \bar{P}_1^k(t) + (M - n_k) \cdot \bar{P}_2^k(t) \tag{18}$$

where

$$\bar{P}_1^k(t) = \int_{\beta_1}^{\beta_2} \int_{\bar{D}_1}^{\bar{D}_2} \{1 - R(t;\text{repair} \mid \beta)\}\bar{D} \cdot p^k(\beta, \bar{D})\, \mathrm{d}\bar{D}\, \mathrm{d}\beta \tag{19}$$

$$\bar{P}_2^k(t) = \int_{\beta_1}^{\beta_2} \int_{\bar{D}_1}^{\bar{D}_2} \{1 - R(t;\text{no repair} \mid \beta, \bar{D})\}\bar{D} \cdot p^k(\beta, \bar{D})\, \mathrm{d}\bar{D}\, \mathrm{d}\beta \tag{20}$$

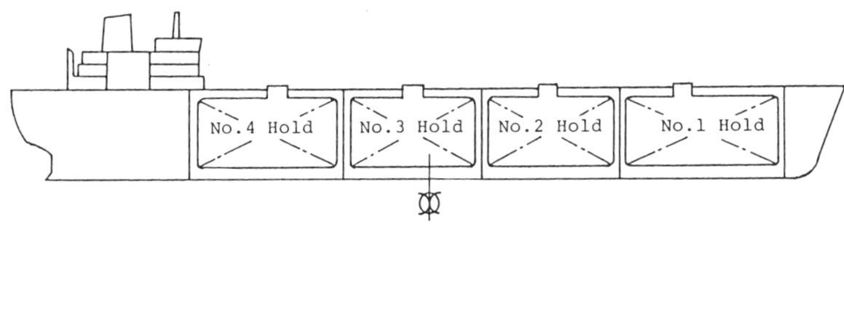

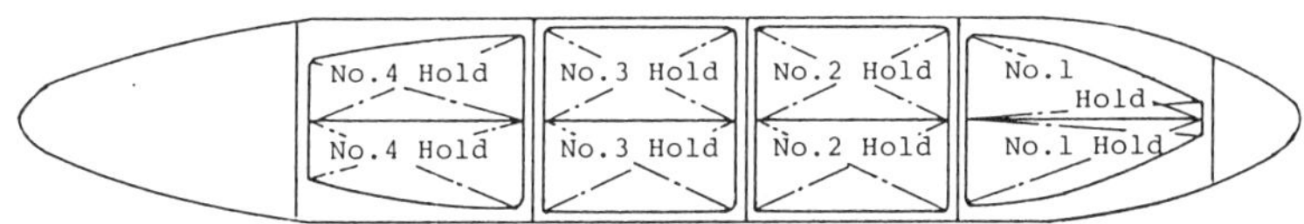

FIG. 1. Schematic drawing of the sample ship.

Comparison with Field Data

By comparing with data obtained from inspections of the lower ends of the hold frames in a ship's structure, the capability of estimating the number of cracked members was investigated to demonstrate the validity of the analytical method just mentioned.

The ship adopted in this analysis was an LPG tanker constructed in 1966 (Fig. 1). This ship has been in regular service with the North Pacific Ocean Line.

Table 1 shows the inspection results of fatigue cracks detected visually and the subsequent repairs to this ship. The situation of fatigue cracks emanating in members of No. 1 and No. 4 holds is different from that in the other holds. Therefore, one hundred members in No. 2 and No. 3 holds are accepted to belong to the same population. Figure 2 depicts the configuration of the frames and of typical web cracks initiated in the frames.

Results and Discussions

Employing the frequencies in Table 1, three kinds of cumulative frequency are plotted in Fig. 3 in order to determine the shape parameter α in Eqn. (1). In consequence, it can be shown that $\alpha = 1$ is appropriate. Several case studies on the range of β and $\bar{D}$ were then performed.

A typical example is depicted in Fig. 4. Table 2 shows the parameter

TABLE 1
Inspection results for detected cracks and subsequent repairs to the lower end of ordinary hold frames

Time after construction		*Number of detected cracked members*															
		No. 1 Hold				*No. 2 Hold*				*No. 3 Hold*				*No. 4 Hold*			
Year	*Month*	*Original member*		*Repaired member*		*Original member*		*Repaired member*		*Original member*		*Repaired member*		*Original member*		*Repaired member*	
		(P)	(S)	(P)	(S)	(P)	(S)	(P)	(S)	(P)	(S)	(P)	(S)	(P)	(S)	(P)	(S)
1	4	8	6	—	—	0	0	—	—	0	0	—	—	0	0	—	—
2	8	0	0	8	5	0	0	0	0	0	0	0	0	0	0	0	0
4	4	13	4	0	0	0	0	0	0	8	7	0	0	0	0	0	0
4	7	0	3	0	0	0	2	0	0	6	5	0	0	0	0	0	0
6	10	1	1	1	0	7	3	0	0	0	0	3	1	1	0	0	0
7	8	1	0	0	0	1	4	3	2	0	0	0	0	0	0	0	0
8	3	0	1	0	0	3	2	0	3	0	0	0	1	0	0	0	0
10	6	2	1	0	1	1	2	5	4	1	0	1	0	0	0	0	0
11	7	0	0	0	2	2	0	2	1	2	1	1	0	1	0	0	0
12	8	0	0	0	1	2	2	3	1	0	2	2	0	0	0	0	0
13	8	2	0	2	0	0	2	3	5	1	3	2	0	0	1	0	0
14	8	0	0	0	0	2	0	1	1	1	1	0	0	0	0	1	0
15	9	0	0	0	0	1	0	1	0	0	0	2	1	0	1	0	0

P: Port side, S: Starboard side.

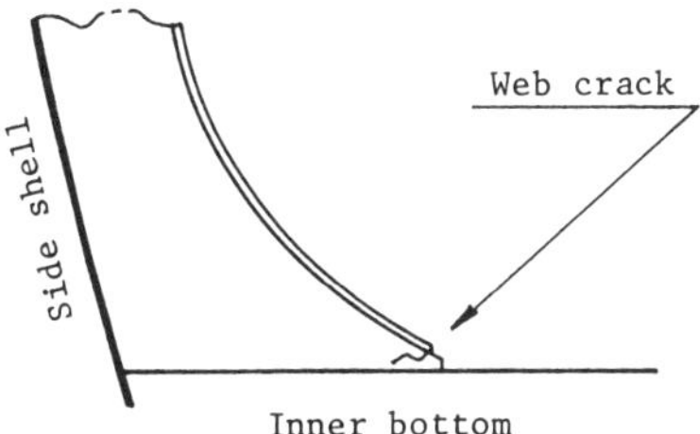

FIG. 2. Structural type of the lower end of the frame and a typical example of the initiated crack.

values of β_1, β_2, $\bar{D}_1$, and $\bar{D}_2$ for the initial prior PDF, in which solid circles depict actual data on the number of cracked members detected. The estimated numbers of cracked members detected, which were calculated by Eqn. (18), are also plotted by symbols in this figure. During several early inspections, the difference between the actual data and the estimate was extreme because fatigue cracks had not

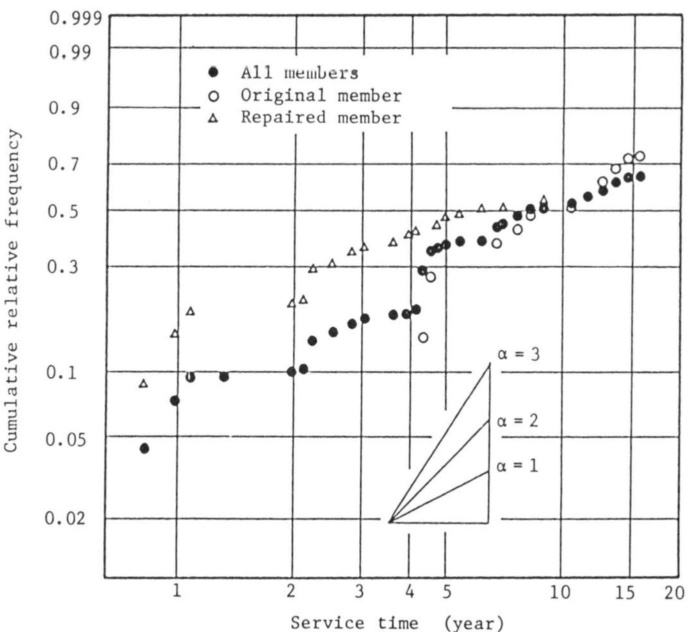

FIG. 3. Relationship between the service time and cumulative frequency.

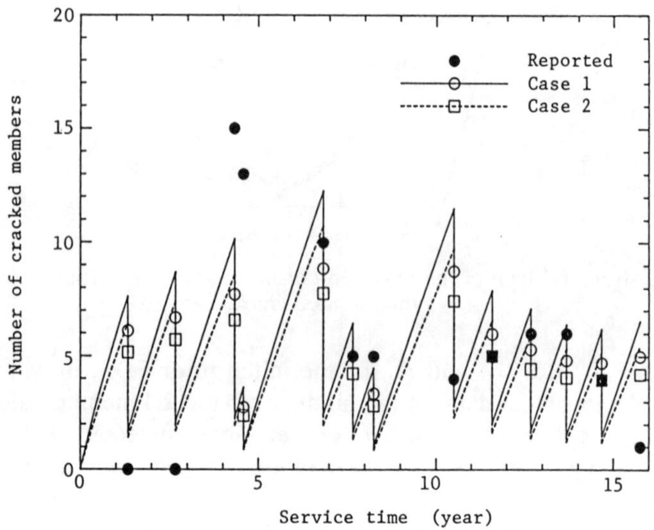

FIG. 4. Relationship between the service time and Bayesian expected number of cracked members.

been detected at the first and the second inspections. However, as many fatigue cracks were detected at the third and the fourth inspections, both estimations for the number of cracked members detected approached the actual data. Both lines in this figure depict the estimated number of cracked members calculated by Eqn. (15). Peaks of the lines indicate the number of cracked members just before the kth inspection, which were estimated by the $(k-1)$th inspection

TABLE 2
Parameter values used to investigate the effect of the prior value of scale parameter for fatigue life

Shape parameter for fatigue life: α	1·0	
Scale parameter for fatigue life	[Case 1]	[Case 2]
Lower limit: β_1	9·2	11·7
Upper limit: β_2	20·8	25·0
Average detectability		
Lower limit: $\bar{D}_1$	0·7	
Upper limit: $\bar{D}_2$	0·9	

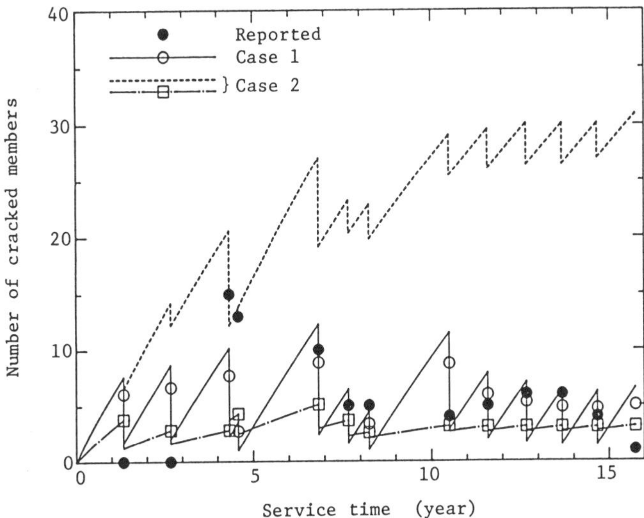

FIG. 5. Effect of the prior value of average detectability.

results. On the other hand, troughs indicate the number of remaining cracked members not detected at the kth inspection, which were estimated by the kth inspection results.

Figure 5 shows the results of investigating the effect of prior average detectability. When the range listed in Table 3 is selected as much wider, the capability of estimation is decreased; that is, the expected number of cracked members shown by the dotted line is increased markedly. On the contrary, the number of cracked members detected as shown by the chain line is decreased.

TABLE 3
Parameter values used to investigate the effect of the prior value of average detectability

Shape parameter for fatigue life: α	1·0	
Scale parameter for fatigue life		
Lower limit: β_1	9·2	
Upper limit: β_2	20·8	
Average detectability	[Case 1]	[Case 2]
Lower limit: $\bar{D}_1$	0·7	0·1
Upper limit: $\bar{D}_2$	0·9	0·9

PROPER INSPECTION INTERVAL BASED ON THE BAYESIAN METHOD

It is significant to establish a flexible inspection program whose inspection interval can be modified in response to service experience to ensure continuing structural integrity. Although the inspection programs adopted by operators at present, as a matter of course, successively reflect the service experience already mentioned and are improved properly, it is pointed out that systematic methods, which can reasonably modify existing inspection programs, have not been sufficiently investigated.

Consequently, the next example [14] discusses an inspection program which can successively modify the inspection intervals for future inspection with the aid of the Bayesian method to estimate uncertain factors and parameters using field data collected during in-service inspection. The inspection interval is evaluated by the criterion that target structures can maintain a predetermined level of structural reliability in order to prevent fatigue failure.

Analytical Model and Formulation

Analytical Model

The following assumptions are acceptable for fatigue crack initiation and propagation, the probability of crack detection and the definition of failure in this analysis.

Fatigue crack initiation. It is assumed that the time to crack initiation t_0 is governed by the two-parameter Weibull distribution given by Eqn. (1). A scale parameter β, which is considered uncertain, can be estimated by the Bayesian method as in the former example.

The crack length initiated at t_0 is x_1, which corresponds to the minimum detectable crack length.

The probability of fatigue crack initiation within the time interval $[(k-1)\cdot\Delta t, k\cdot\Delta t]$ is given by

$$\Delta F(k) = \exp[-\{(k-1)\cdot\Delta t/\beta\}^{\alpha}] - \exp[-(k\cdot\Delta t/\beta)^{\alpha}] \qquad (21)$$

and the probability of no crack initiation by the time $k\cdot\Delta t$ is

$$R_F(k) = \exp[-(k\cdot\Delta t/\beta)^{\alpha}] \qquad (22)$$

in which Δt is a unit time interval.

Fatigue crack propagation. Simulation of the fatigue crack propagation was conducted by applying the Markov chain model proposed by Bogdanoff [15]. The procedure to calculate crack propagation involves a transition matrix T to define the Markov chain, which can be described as follows:

$$\boldsymbol{T} = \begin{pmatrix} u_1 & v_1 & 0 & 0 & \cdots & 0 & 0 & 0 \\ 0 & u_2 & v_2 & 0 & \cdots & 0 & 0 & 0 \\ 0 & 0 & u_3 & v_3 & \cdots & 0 & 0 & 0 \\ \vdots & \vdots & \vdots & \vdots & \cdots & \vdots & \vdots & \vdots \\ 0 & 0 & 0 & 0 & \cdots & 0 & u_{d-1} & v_{d-1} \\ 0 & 0 & 0 & 0 & \cdots & 0 & 0 & 1 \end{pmatrix} \tag{23}$$

with

$$u_k + v_k = 1 \qquad k = 1, 2, \ldots, d \tag{24}$$

in which u_k is the probability that a crack of length x is in a state k, and v_k is the transition probability that a crack of length x_k in a state k is propagated to the next state $(k+1)$ after a unit time interval Δt. When a homogeneous Markov chain is considered, the following recurrent relationship between the row vectors of crack distribution $\mathbf{A}_j$ at the time $j \cdot \Delta t$ and $\mathbf{A}_{j-1}$ at $(j-1) \cdot \Delta t$ can be derived by using the matrix $\boldsymbol{T}$.

$$\mathbf{A}_j = \mathbf{A}_{j-1} \cdot \boldsymbol{T} \tag{25}$$

in which $\mathbf{A}_j$ is defined by

$$\mathbf{A}_j = \{p_j(x_1), p_j(x_2) \cdots p_j(x_d)\} \tag{26}$$

and the element $p_j(x_k)$ is the probability of the crack length being x_k. As x_d is the maximum crack length being considered, a fatigue crack would not be propagated over this length in the present analysis.

Probability of crack detection. The probability of detecting a crack of length x is defined by the following equation with parameters a_1, a_2 and m:

$$\begin{aligned} D(x) &= 0 & a_1 > x \\ &= \{(x-a_1)/(a_2-a_1)\}^m & a_2 \geq x \geq a_1 \\ &= 1 & a_2 < x \end{aligned} \tag{27}$$

Therefore, the probability of not detecting a crack of length x becomes

$$D^*(x) = 1 - D(x) \tag{28}$$

Repair. When a fatigue crack is detected, it is repaired or replaced to reproduce the initial state.

Definition of failure. As already mentioned, fatigue cracks are arrested at the maximum length x_d, which corresponds to the so-called fail-safe crack length, and detected cracks are repaired prior to being subjected to the critical load. Therefore, the probability of failure is defined by $p_j(x_d)$ of Eqn. (26) in this analysis.

Formulation

When one critical element in a structure is considered and a fatigue crack emanates under the condition of the time to crack initiation governed by Eqn. (1), the probability of the outcomes caused by an inspection can be obtained in accordance with the following procedure.

The first inspection. The row vector of the crack distribution at every time interval Δt can be described as

$$\left.\begin{aligned}
&\mathbf{A}_1 = F_1 = \{p_1(x_1), 0, 0 \cdots 0\}\\
&\mathbf{A}_2 = \mathbf{F}_2 + \mathbf{A}_1 \cdot \boldsymbol{T} = \{p_2(x_1), p_2(x_2), 0, 0 \cdots 0\}\\
&\vdots\\
&\mathbf{A}_n = \mathbf{F}_n + \mathbf{A}_{n-1} \cdot \boldsymbol{T} = \{p_n(x_1), p_n(x_2) \cdots p_n(x_n), 0, 0 \cdots 0\}\\
&\vdots\\
&\mathbf{A}_{j_1} = \mathbf{F}_{j_1} + \mathbf{A}_{j_1-1} \cdot \boldsymbol{T} = \{p_{j_1}(x_1), p_{j_1}(x_2) \cdots p_{j_1}(x_\mathrm{d})\} \quad j_1 \geq d
\end{aligned}\right\} \tag{29}$$

in which, using Eqn. (21), $\mathbf{F}_k$ becomes

$$\mathbf{F}_n = \{\Delta F(n), 0, 0 \cdots 0\} \qquad n = 1, 2, 3, \cdots, j_1 - 1, j_1 \tag{30}$$

When the first inspection has been performed at the time $j_1 \cdot \Delta t$ mentioned above, the probabilities of crack detection and no crack detection are given by

$$P_\mathrm{D}(1) = \sum_{\mathrm{k}=1}^{\mathrm{d}} p_{j_1}(x_k) \cdot D(x_k) \tag{31}$$

$$P_\mathrm{D}^*(1) = \sum_{k=1}^{d} p_{j_1}(x_k) \cdot D^*(x_k) + R_\mathrm{F}(1) \tag{32}$$

Thus, the row vector $\mathbf{A}_{j_i}^*$ of the remaining crack distribution just after the first inspection is

$$\mathbf{A}_{j_1}^* = \{p_{j_1}(x_1) \cdot D^*(x_1), p_{j_1}(x_2) \cdot D^*(x_2) \cdots p_{j_1}(x_d) \cdot D^*(x_d)\} \quad (33)$$

The second inspection and the ith inspection. After the first inspection, the row vector at every time interval can be described as

$$\left.\begin{aligned} \mathbf{A}_{j_1+1} &= \mathbf{F}_{j_1+1} + \mathbf{A}_{j_1}^* \cdot \boldsymbol{T} \\ \mathbf{A}_{j_1+2} &= \mathbf{F}_{j_1+2} + \mathbf{A}_{j_1+1} \cdot \boldsymbol{T} \\ &\vdots \\ \mathbf{A}_{j_2} &= \mathbf{F}_{j_2} + \mathbf{A}_{j_2-1} \cdot \boldsymbol{T} \end{aligned}\right\} \quad (34)$$

The second inspection is performed at the time $j_2 \cdot \Delta t$.

In a similar fashion, the probabilities of crack detection and no crack detection after the ith inspection are given by

$$P_{\mathrm{D}}(i) = \sum_{k=1}^{d} p_{j_i}(x_k) \cdot D(x_k) \quad (35)$$

$$P_{\mathrm{D}}^*(i) = \sum_{k=1}^{d} p_{j_i}(x_k) \cdot D^*(x_k) + R_{\mathrm{F}}(i) \quad (36)$$

The row vector $\mathbf{A}_{j_i}^*$ of remaining crack distribution is

$$\mathbf{A}_{j_i}^* = \{p_{j_i}(x_1) \cdot D^*(x_1), p_{j_i}(x_2) \cdot D^*(x_2) \cdots p_{j_i}(x_d) \cdot D^*(x_d)\} \quad (37)$$

Outcomes of an inspection and their probability. With regard to a critical element l in a structure, the probability of detecting a crack in the element at the ith inspection with the time record that a crack has been detected and repaired S times can be calculated from Eqn. (35) as follows:

$$\begin{aligned} P_{\mathrm{E}}(i:l) = P_{\mathrm{D}}(\tau_1; l) \cdot P_{\mathrm{D}}(\tau_2; l) \cdots P_{\mathrm{D}}(\tau_s; l) \\ \cdot P_{\mathrm{D}}(\tau_{s+1} = (j_i - j_s) \cdot \Delta t; l) \end{aligned} \quad (38)$$

in which τ_n ($n = 1, 2, \ldots, s, s+1$) is the time interval between the gth inspection that detects a crack in an element and the hth inspection which comes after the gth and detects a crack in the same element again, and is defined as

$$\tau_n = (j_g - j_h) \cdot \Delta t \quad (39)$$

When a crack is not detected at the ith inspection under the same time history as that just mentioned, the probability of not detecting a crack is

$$P_{\mathrm{E}}(i; l) = P_{\mathrm{D}}(\tau_1; l) \cdot P_{\mathrm{D}}(\tau_2; l) \cdots P_{\mathrm{D}}(\tau_s; l) \cdot P_{\mathrm{D}}^*(\tau_{s+1}; l) \qquad (40)$$

Accordingly, in the case of a fleet size L, the probability of the outcomes caused by an inspection becomes

$$P_{\mathrm{EF}}(i) = \prod_{l=1}^{L} P_{\mathrm{E}}(i; l) \qquad (41)$$

Bayesian Analysis

The uncertain scale parameter β in the PDF of fatigue crack initiation is estimated by the same method as that mentioned in the preceding section; namely, the initial prior PDF, $p^0(\beta)$, of β is selected to be uniformly distributed over the range of β_1 and β_2. By replacing $P_{\mathrm{EF}}(i)$ in Eqn. (41) with the conditional probability $P_{\mathrm{EF}}(i \mid \beta)$ of β, the posterior PDF of β after the ith inspection has been completed can be obtained from

$$p^i(\beta) = P_{\mathrm{EF}}(i \mid \beta) \cdot p^0(\beta) \Big/ \int_{\beta_1}^{\beta_2} (\text{Numerator})\, \mathrm{d}\beta \qquad (42)$$

As mentioned in the assumptions, it has been defined that a failure of the element occurs if a crack in the element is propagated to the fail-safe crack length x_d. Therefore, using the remaining crack distribution of Eqn. (37), the probability of failure of the lth structure just before the $(i+1)$th inspection is

$$P_{\mathrm{F}}(i+1; l \mid \beta) = P_{j_{i+1}}(x_d) \qquad (43)$$

and the probability of the first failure in the fleet is given by

$$P_{\mathrm{FF}}(i+1 \mid \beta) = 1 - \prod_{l=1}^{L} \{1 - P_{\mathrm{F}}(i+1; l \mid \beta)\} \qquad (44)$$

An expected probability of failure can be obtained employing Eqns. (42) and (44) as

$$\bar{P}_{\mathrm{FF}}(i+1) = \int_{\beta_1}^{\beta_2} P_{\mathrm{FF}}(i+1 \mid \beta) \cdot p^i(\beta)\, \mathrm{d}\beta \qquad (45)$$

Determination of the Inspection Interval

The $(i+1)$th inspection should be conducted to satisfy the condition that the expected probability of failure P_{FF} shown by Eqn. (45) is less than the given allowable probability of failure for a fleet $P_{FFC}(i+1)$. This safety criterion can be expressed as follows:

$$\bar{P}_{FF}(i+1) \leq P_{FFC}(i+1) \tag{46}$$

Consequently, the time interval between the ith and the $(i+1)$th inspection is defined by

$$(j_{i+1} - j_1) \cdot \Delta t \leq \bar{P}_{FF}^{-1}\{P_{FFC}(i+1)\} \tag{47}$$

Numerical Example

Parameter values chosen for the numerical examples are shown in Table 4, in which the time in the model has been normalized. Simulations were performed on three cases for the true value β (Case 1: 0·75, Case 2: 1·0, Case 3: 1·25). Assuming the power law for crack

TABLE 4
List of parameter values

Fatigue crack initiation			
Shape parameter: α		4·0	
Scale parameter: β	Case 1	Case 2	Case 3
True value (unknown)	0·75	1·0	1·25
Prior estimation of β			
Lower limit: β_1		0·5	
Upper limit: β_2		1·5	
Fatigue crack propagation			
Initial crack length: x_1		0·5	
Fail-safe crack length: x_d		5·0	
Mean of crack propagation period: ΔT		0·2	
$\sigma_t^2/\bar{t}$: r		5·0	
NDI detectability			
m		0·2	
a_1		0·5	
a_2		5·5	
Unit time interval: Δt		0·002	
Allowable probability of			
Failure for a fleet: P_{FFC}		0·01	
Fleet reliability: $1 - P_{FFC}$		0·99	
Fleet size: L		30	

TABLE 5
Mean of the crack propagation period

$x(t)$	$\bar{t}$
$0{\cdot}5(x_1)$	0
1·0	0·0857
2·0	0·1463
3·0	0·1731
4·0	0·1891
$5{\cdot}0(x_d)$	0·2 (ΔT)

Note: $\bar{t} = Q \cdot \{x(t)^{-0{\cdot}5} - x_1^{-0{\cdot}5}\}$, $Q = -0{\cdot}20683$.

propagation with the parameters in Table 4 and the mean of the crack propagation period $\bar{t}$ in Table 5, the elements of the transition matrix can be calculated from the relationship that the ratio $r = \sigma_t^2/t$ is constant during the crack propagation period [16], in which σ_t^2 is the variance of the crack propagation period. The element values obtained are shown in Table 6, and some samples of crack propagation and their detectability are depicted in Fig. 6. The fleet reliability is required to be constant during the service time in this example, and its value is 0·99. In other words, the allowable probability of failure for a fleet is required to be less than $P_{\text{FFC}} = 0{\cdot}01$.

TABLE 6
Elements of the transition matrix

k	r_k	u_k	v_k
1 ~ 14	5·119	0·8366	0·1634
15 ~ 20	5·329	0·8419	0·1581
21 ~ 24	4·658	0·8232	0·1768
25 ~ 27	4·146	0·8057	0·1943
28 ~ 29	4·695	0·8244	0·1756
30 ~ 31	3·340	0·7743	0·2257
32	6·140	0·8599	0·1401
33	4·910	0·8308	0·1692
34	4·010	0·8004	0·1996
35	—	1·0	—

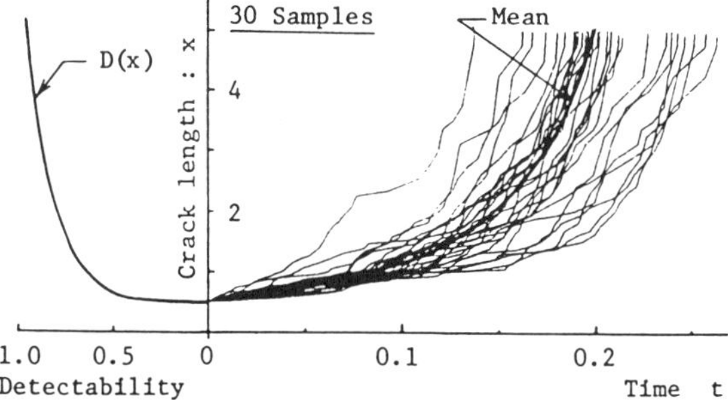

FIG. 6. Samples of crack propagation generated by a Markov chain model and NDI detectability.

Results and Discussion

Examples of the inspection schedule are given in Fig. 7 with the condition for continuing the constant reliability. The relationships between the estimated inspection schedule and probability of failure are drawn by the solid lines, while the dotted lines represent the results for the true value of β. The dots along the abscissa indicate the execution times for the inspections. As can be seen from this result, the inspection intervals are different from each other and especially are shorter in the proximity of the true value of β because of the safety criterion of constant fleet reliability. In order to satisfy the safety criterion, the required numbers of inspections for Cases 1–3 performed up to the service time $t = 1{\cdot}5$ are 24, 23 and 22, respectively. It can be seen that the estimated inspection time for Case 1 is slightly unconservative at the early stage, and that of Case 3, on the contrary, is conservative. This fact shows that inspection schedules thus determined are dependent on the range of the initial prior density of β. However, when the service time reaches about 50% of the true value of β, the uncertain factor β can be estimated successfully in the case of the fleet size in this example.

Figure 8 depicts the example of the inspection interval between the ith inspection and the $(i-1)$th inspection. The estimated inspection intervals equivalent to the dot intervals in Fig. 6 are shown by the solid lines. The dotted lines are the inspection intervals for the true β calculated by the same safety criterion as that used for estimating the inspection program. Accordingly, it is pointed out that both inspection intervals are almost the same except for the early stage when the inspections were started.

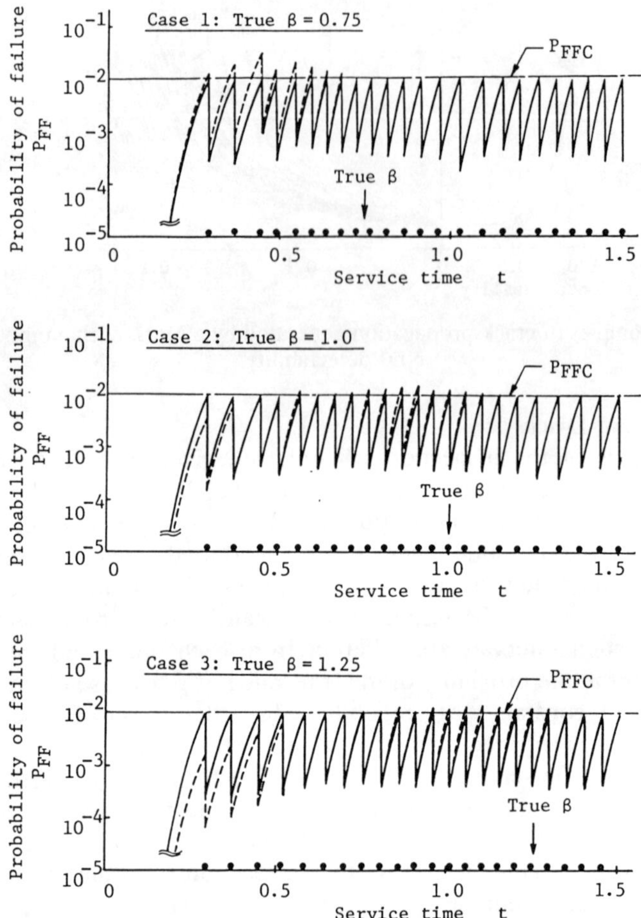

FIG. 7. Relationship between the estimated inspection interval and probability of failure.

CONCLUSIONS

Two typical examples have been described from several investigations to demonstrate the validity of Bayesian reliability analysis in meeting the requirement for estimating and updating uncertain factors with the aid of field data gathered during in-service inspections. One presents an analysis for a ship's structure with uncertain factors of the

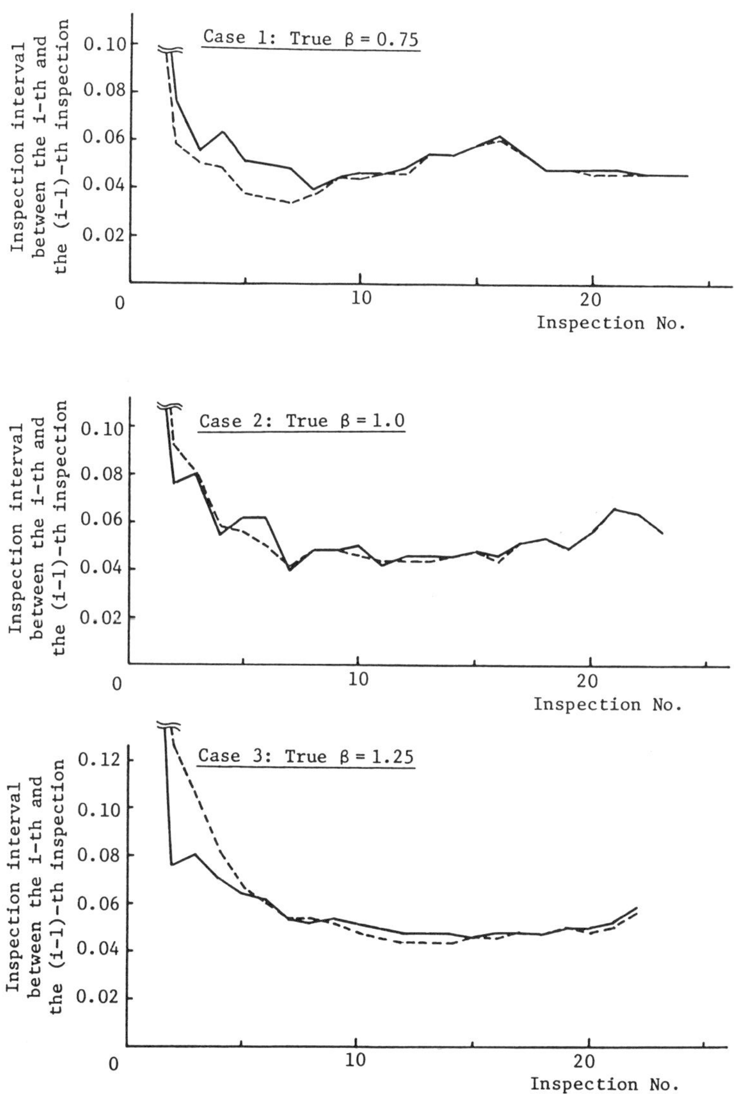

FIG. 8. Relationship between the inspection time and inspection interval.

parameter in the PDF of fatigue life and NDI detectability. As already mentioned, it is significant to note that actual field data for the lower end of frames were applied in this example. The other example discusses the determination of a proper inspection interval after each in-service inspection for estimating the parameter in the PDF of fatigue life. Both examples indicate that uncertain factors have been satisfactorily estimated after several early in-service inspections.

According to the results of other numerical examples involving the estimation of parameters in the PDF from fatigue test results, it is also pointed out that a Bayesian analysis is effective in appropriately estimating uncertain parameters. Therefore, further investigations should be conducted on actual in-service structures with field data.

ACKNOWLEDGEMENTS

The authors wish to express their sincere appreciation to Prof. M. Shinozuka of Columbia University and Mr A. Nitta of Nippon Kaiji Kyokai for their valuable suggestions and encouragement during this study, and also thank Dr H. Asada of the National Aerospace Laboratory for his assistance in preparing this report.

REFERENCES

[1] J.-N. Yang and W. J. Trapp, *AIAA J.*, **12,** 1623 (1974).

[2] M. Shinozuka, *AFFDL-TR-76-31* (1976).

[3] C. A. Cornell, *ICOSSAR '69,* A. M. Freudenthal, Ed., Pergamon Press, Oxford, 47 (1972).

[4] H. Itagaki, F. Ozaki and T. Nemoto, *J. Soc. Naval Arch. Jap.*, **138,** 447 (1975).

[5] H. Itagaki, F. Ozaki and T. Nemoto, *J. Soc. Naval Arch. Jap.*, **139,** 307 (1976).

[6] H. Itagaki, T. Ogawa and S. Yamamoto, *J. Soc. Naval Arch. Jap.*, **141,** 263 (1977).

[7] H. Itagaki and H. Asada, *HOPE International JSME Symp.*, Tokyo, 481 (1977).

[8] M. Shinozuka, H. Itagaki and H. Asada, *Proc. US–Japan Cooperative Seminar on Fracture Tolerance Evaluation,* K. Kanazawa, A. S. Kobayashi and K. Iida, Eds., Hawaii, USA, 237 (1981).

[9] H. Itagaki, H. Ishikawa and N. Yamamoto, *J. Soc. Naval Arch. Jap.*, **153,** 376 (1983).

[10] H. Itagaki, Y. Akita and A. Nitta, *Proc. Role of Design, Inspection and Redundancy in Marine Structural Reliability,* Williamsburg, Virginia, USA, 71 (1983).
[11] H. Itagaki and N. Yamamoto, *Proc. ICOSSAR '85,* I. Konishi, A. H-S. Ang and M. Shinozuka, Eds., IASSAR, Kobe, Japan, **3,** 533 (1985).
[12] H. Asada, H. Itagaki and S. Itoh, *Proc. ICOSSAR '85,* I. Konishi, A. H-S. Ang and M. Shinozuka, Eds., IASSAR, Kobe, Japan, **1,** 87 (1985).
[13] H. Itagaki and N. Yamamoto, *J. Soc. Naval Arch. Jap.,* **158,** 565 (1985).
[14] H. Itagaki and S. Itoh, *Proc. International Conference on Fracture and Fracture Mechanics,* Fudan University Press, Shanghai, China, 391 (1987).
[15] J. L. Bogdanoff, *ASME J. Appl. Mech.,* **45,** 246 (1978).
[16] Y. Shimada, T. Nakagawa and H. Tokunou, *J. Soc. Mater. Sci. Jap.,* **33,** 367 (1974).

[illegible]

[11] [illegible]

[12] [illegible]

[13] [illegible] and N. Yamamoto, [illegible], 368, [illegible]

[14] H. [illegible] and [illegible], Proc. International Symposium on [illegible] and [illegible], [illegible] University, [illegible] 1988.

[15] [illegible], Scripta Metall., 18, [illegible]

[16] [illegible] and H. [illegible], [illegible], 32, [illegible] 1984.

Expectations from Knowledge Engineering for Reliability Improvement

SHUICHI FUKUDA

Welding Research Institute, Osaka University, Ibaraki, Osaka, Japan

ABSTRACT

This paper demonstrates that knowledge engineering offers us much more as a debugger or tracer of knowledge than a tool for symbolic processing in terms of reliability analysis/synthesis. Knowledge engineering provides us with the computer environment to follow very different reasoning processes or to look at things from different perspectives. Thus it enables us to become aware of shortcomings in our knowledge or to discover what has been overlooked. Therefore, we should not consider knowledge engineering just as one of these conventional tools just to reach the goal in a very short and effective manner but rather we should use it as a tool to examine the processes how to reach there.

INTRODUCTION

This paper discusses what we can expect from knowledge engineering to improve reliability, using experience from the development of such expert systems as a design and maintenance support system for oil storage tanks, and an advisory system for welding.

There have been many papers on the application of knowledge engineering to reliability problems like diagnostics, and the usefulness of knowledge engineering in this field seems to have already been established.

However, most of these papers draw our attention toward the point that knowledge engineering can process symbolic knowledge with relative ease, which has been very difficult with conventional programming, although it is very important from the standpoint of reliability analysis, and that knowledge engineering can deal with combinatorics problems to a much greater degree than otherwise.

It is not the intention of the author to repeat these discussions here, but to discuss knowledge which seems to be very different from engineer to engineer, especially if their experience is different. In other words, an engineer cannot be convinced of the results unless his or her experience is truly reflected in the outcome.

We will point out that the knowledge engineering approach more directly reflects the reasoning processes of engineers so that, even if the goal is identical or is bounded by a small region, the ways to reach the goal differ very widely from engineer to engineer according to their engineering background or experience. This characteristic is expected to help a great deal in improving reliability, although this point seems to have been little discussed until now.

KNOWLEDGE ENGINEERING AND THE CONVENTIONAL APPROACH: WHAT IS THE DIFFERENCE?

It is believed that the outcome of an engineering analysis does not differ from engineer to engineer. Although this is true in most cases, is it true if we consider a very complicated problem, and especially an ill-structured one?

The main difference between the knowledge engineering approach and the conventional one is that knowledge engineering is more process-oriented while the conventional approach is more goal-oriented. In other words, the way(s) to arrive at the conclusion is definitely predetermined with conventional programming, while the knowledge engineering approach processes the data as they come in so that the reasoning process is directly reflected in the sessions. Especially if the problem is ill-structured, the ways to direct the problem toward its goal differ widely from engineer to engineer.

Let us take the example of the WELCON development [1], which is a system to provide advice on the determination of welding conditions. As welding is related to a wide variety of engineering fields from mechanical to electrical and metallurgical, the reasoning process differs from engineer to engineer with their different engineering backgrounds, although the final conclusions fall within narrow bounds. It should be stressed that an engineer is not convinced of the conclusion unless the reasoning process is similar to his or her own.

It is quite interesting to note that engineers do not agree easily if the reasoning process is different from their own, even if the pieces of knowledge they offer for representation are identical or at least do not differ appreciably. The difference in the reasoning process comes from how these pieces of knowledge are assembled or when to apply them.

The greatest disagreement can be seen among engineers when it comes to the utilization of their experience. Experience is of course a very personal matter, so that it is difficult to reproduce it on a computer. To make it possible, we have first to reproduce the reasoning process of each engineer and make efforts to put matters in his or her perspective, because the difference of perspective is most apparently reflected by the reproduction of experience. A similar situation will be experienced differently by different engineers, and to reproduce this situation, it is necessary to look at it from the viewpoint of each engineer.

This problem was also apparent when we developed DEST-I [2] and MAST-I [3], the design and maintenance support systems, respectively, for oil storage tanks. In the case of design, the final outcomes were very different from designer to designer, so that the situations were very different from the WELCON case in which the final results fell within narrow bounds; but it should be stressed that identical or similar pieces of knowledge will be applied quite differently from designer to designer according to experience, which is quite similar to the case we encountered during the WELCON development.

The MAST-I development involved situations that were somewhat different, because the main target of MAST-I is the maintenance of a structure against corrosion. In the field of corrosion, the same phenomenon is known by a different name by researchers with different backgrounds, although the foregoing observations fundamentally hold true in this case.

The fact that such experiences were obtained in the development of expert systems is because knowledge engineering is fundamentally data-driven, while conventional programming is procedure-driven. Although the same data are treated differently from engineer to engineer, conventional programming allows only a few ways, at most, of processing them.

Up to now, it has been widely accepted among engineers that computers produce identical or similar outcomes, and the route taken has not received much attention. However, with design for example, the final design will differ quite widely from designer to designer, so that the conventional programming approach is not fitted for the purpose, although knowledge engineering is a strong candidate.

Even if the final outcome falls within narrow bounds, the route taken will vary very much from engineer to engineer, although not much attention has been paid to the problem.

Knowledge engineering as it is accepted today is more emphasized as a tool for symbolic processing, but it can be used as a reliability analysis/synthesis tool as well.

KNOWLEDGE ENGINEERING AS A DEBUGGER OR TRACER OF KNOWLEDGE

Although knowledge engineering is data-driven, most engineers realize its value based on the outcomes of its analyses, which has been the custom with conventional programming. This is also sometimes true when it is used as a reliability analysis/synthesis tool. If we introduce symbolic processing, a considerable reduction of programming effort can be expected since quite a large proportion of reliability knowledge is composed of symbols, which seem to be involved more than numerals, because symbols can be applied more generally than numerals.

Let us now consider the case of math study. Textbooks show how to solve the problems, but we do not learn to solve math problems just by following these textbook procedures. It is, in fact, the other way around, because we make many mistakes in solving a math problem and reaching the final solution, repeating this process many times until we learn the way to solve the problem. Therefore, the way to solve the same problem differs from person to person because the mistakes made also differ. The textbook shows the shortest and most elegant way to the solution, and does not necessarily correspond to the one each student will follow.

It has been pointed out that more than 80% of accidents could be foreseen and prevented [4]. Such disasters, in most cases, are due to overlooking the foreseeable factors, which occurs because most people look at things from their own perspective and are very reluctant to look at them in any other way. Each engineer has a particular frame of mind and what comes within this is processed very quickly and effectively, although anything outside is more often than not neglected or overlooked.

Knowledge engineering can provide the means for processing data according to each engineer's own methods if the reasoning process of each engineer can be traced and implemented on a computer. If such traces of the reasoning process are stored in a large enough quantity, then it would become possible for each engineer to view things in a perspective different from his or her own so that such overlooking could possibly be prevented.

This would produce another effect of providing a better communication tool. An engineer can look at things from another point of view and thus he knows what they have in common and how they differ. Therefore, the differences could be identified and the points argued better.

Another advantage of knowledge engineering in addition to this tracing role is that of debugging. We have many shortcomings in our knowledge, but are not aware of them until we come across a shortage of knowledge in a real incident. It is often pointed out that, in most cases, the knowledge to prevent an accident is already known to somebody else. The cause of a disaster is often the lack of a proper tool to transfer knowledge between people.

This is especially true if the problem is ill-structured and has many influencing factors. We cannot put these factors into proper balance, we sometimes overlook important factors and, what is worse, we did not recognize our shortcomings of knowledge in evaluating these factors.

As knowledge engineering permits many different types of processing of the same data, we can take different reasoning processes to debug our knowledge. Thus, we can recognize the shortcomings in our knowledge and rectify the problem.

It should be stressed that the role of a computer is quite different from that in conventional uses, because the final outcomes are not necessarily expected. Although the outcomes in mid-processes can be expected, the debugging role is important and the final outcome follows next.

CONCLUDING REMARKS

Knowledge engineering is process-oriented software in contrast to the conventional goal-oriented programming. Therefore, knowledge engineering can offer more as a debugger or tracer of knowledge than as a tool to reach to the goal in a short and effective manner when we apply it to reliability analysis or synthesis.

Contrary to the common engineering concept that a computer is a machine to produce results in a short and effective way, knowledge engineering offers much more flexibility for processing data, i.e. the number and variety of the reasoning processes the computer generates during a session with a user make it possible for him or her to look at things from a different perspective than usual, or to become aware of shortcomings in knowledge or to discover what has been overlooked.

Thus, it is expected that knowledge engineering will not provide a substitute for conventional programming tools, but rather will compensate for what is currently lacking to help with reliability analysis/synthesis if it is used together with the conventional tools.

REFERENCES

[1] S. Fukuda, Development of an expert system for welding design support, *Computer Applications in Production and Engineering,* K. Bo, L. Estensen, P. Falster and E. A. Warman, Eds., North-Holland, Amsterdam, 679 (1987).

[2] S. Fukuda and T. Motooka, Development of a design support system DEST-I for oil storage tanks, *Computers in Engineering 1985,* Vol. 3, R. Raghaven and S. M. Rohde, Eds., ASME, New York. 441 (1985).

[3] S. Fukuda, Development of a computer based consultation system for preventing structural failures using Prolog, *Proc. 4th Int. Conf. on Structural Safety and Reliability,* Vol. I, IASSAR, Columbia University, New York, 27.

[4] D. L. Marriott and N. R. Miller, Material failure logic models: A procedure for systematic identification of material failure modes in mechanical components, *Failure Prevention and Reliability, 1981,* ASME, New York, 197.

Integrated Color-Face Graphics for Displaying Safety Conditions in an Industrial System

FUMIO HARA

Department of Mechanical Engineering, Science University of Tokyo, Tokyo, Japan

ABSTRACT

An integrated color graphics technique is introduced for displaying system safety conditions, which uses cartoon-like colored illustrations in the form of faces. This is done by drawing the face on a CRT, which involves nonlinearly transforming 31 variables, and then coloring the face. This integrated color graphics is applied to display the safety conditions during the progress of events in the Three Mile Island nuclear power plant accident. Human visual perceptive characteristics are investigated in relation to the perception of the plant accident process, the natural face color change, and the consistency between facial expressions and colors. This paper concludes that colors used in an integrated color-face graphics technique must be completely consistent with the emotional feelings that can be perceived from the colors.

INTRODUCTION

Industrial systems such as chemical plants, power plants and petroleum refineries continue to grow in size and complexity, so that many instrument data, relating not only to plant operation conditions but also to its structural integrity, are needed to safely operate such large-scale systems. However, when these data are displayed on the main control panel, plant operators are liable to have difficulty in properly surveying the plant condition due to the diversity of data displayed on a large panel. This problem becomes much more serious in a plant operation emergency, involving, for example, unknown malfunctions, leakage of chemical liquids from pipelines or abnormal conditions in a reactor. An emergency of this kind may lead to an explosion, fire and/or other catastrophic disasters.

The analysis of industrial plant faults and disasters indicates that methods for communicating information between man as the plant operator and the system of an industrial plant need to be improved if plant faults and disasters are to be prevented.

Clear, correct communication between the operator and system in situations such as those just described in large-scale industrial systems is of paramount importance; first, in ensuring that information on plant conditions is transmitted accurately and, second, in reducing the stress that plant operators experience while monitoring plant conditions. This is absolutely essential in the case of system malfunction or failure, where a misunderstanding or an incorrect judgement of the plant status could cause a catastrophic accident.

What is needed is a man–system communication interface to integrate the large amount of complex plant information and display it so that an operator can understand it easily. Such an interface could reduce the stress that plant operators experience, thus reducing the chance of misjudging plant conditions. From this viewpoint, the discriminative ability of human pattern recognition faculties has recently come under consideration for application to monitoring a plant's condition. Human pattern recognition is, in general, most sensitive to expressions of the human face due to the life-long experience of communication through facial expressions.

This paper describes an integrated communication interface that uses cartoon-like colored illustrations in the form of faces, which, through different facial expressions, can show a plant's safety status. This is done by drawing the face on a CRT by nonlinearly transforming 31 variables and coloring the face based on an integration of these variables. This integrated color graphics technique will then be applied to display the progress of safety conditions in the Three Mile Island (TMI) nuclear power plant accident. The usefulness of the technique will then be discussed in relation to the man-system communication interface that is necessary for industrial system safety.

COLOR–FACE GRAPHIC GENERATION

Face-Graph Drawing

A monochrome face is generated on a CRT screen by organizing four face elements such as the face contour with the hair, nose, mouth and eyes. These elements are eventually drawn using 31 parameters P_1

TABLE 1
Face graph parameters

Face direction	face direction (R–L)	P_1	Expression	mouth shape	P_{10}
	face direction (U–D)	P_2		mouth openness	P_{11}
				mouth inclination	P_{12}
Facial features	face length	P_3		right eye size	P_{14}
	jaw shape	P_4		left eye size	P_{15}
	head shape	P_5		right eye openness	P_{16}
	hair amount	P_6		left eye openness	P_{17}
	nose size	P_7		right eye shape	P_{18}
	nose length	P_8		left eye shape	P_{19}
	mouth size	P_9		right eye inclination	P_{20}
	distance between eyes	P_{13}		left eye incliniation	P_{21}
	right eye position (R–L)	P_{26}		distance between right eye and eyebrow	P_{28}
	left eye position (R–L)	P_{27}		distance between left eye and eyebrow	P_{29}
Eye direction	right eye direction (R–L)	P_{22}		right eyebrow shape	P_{30}
	left eye direction (R–L)	P_{23}		left eyebrow shape	P_{31}
	right eye direction (U–D)	P_{24}			
	left eye direction (U–D)	P_{25}			

to P_{31} indicated in Table 1. These parameters, whose value ranges from 0 to 1, are used to specify the shape of each facial element.

Figures 1(a) to (d) show the face variables necessary to draw a face graph on a CRT screen. For a given location of the center point $F = (F_x, F_y)$ of the face graph, and the half height of a face (F_s), the face contour can be drawn using the following mathematical formula for the lower half of the face graph:

$$\begin{aligned} X_{f1}/F_s &= 1 - u^{C_f} \\ Y_{f1}/F_s &= 1{\cdot}25\{1 - (1 - u^{C_f})\}F_l \\ X_{f0}/F_s &= 0{\cdot}5\{1 - (1 - u)^{C_f}\}^2(P_1 - 0{\cdot}5) \end{aligned} \tag{1}$$

where

$$F_l = 0{\cdot}4P_3 + 0{\cdot}8 \quad \text{and} \quad C_f = 1{\cdot}9P_4^{2{\cdot}25} + 1{\cdot}6 \tag{2}$$

For the upper half of the face graph,

$$\begin{aligned} X_{f2}/F_s &= 1 - u^{H_f} \\ Y_{f2}/F_s &= 1 - (1 - u)^{H_f} \end{aligned} \tag{3}$$

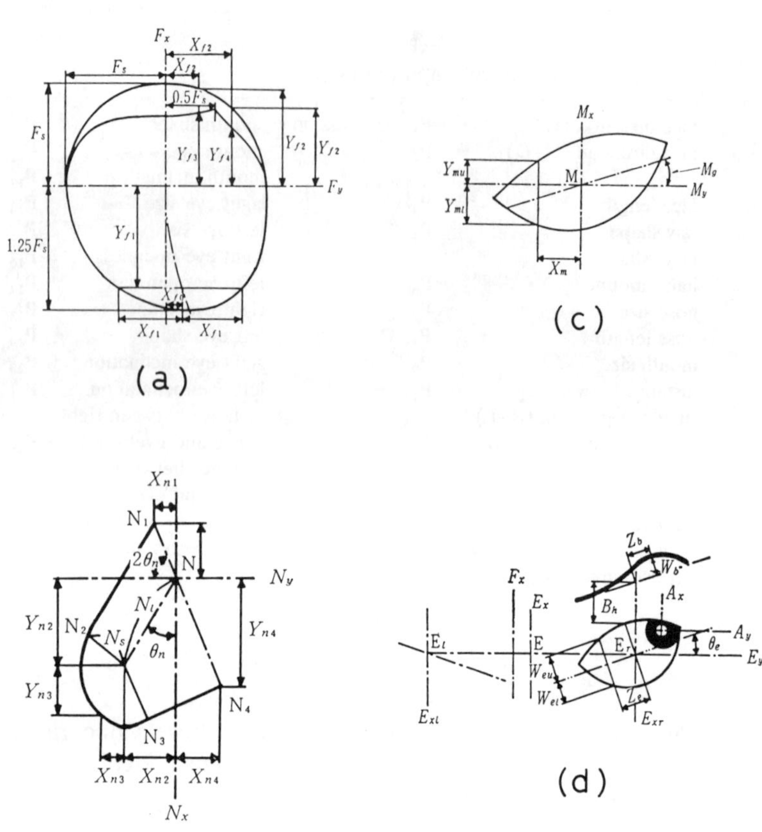

FIG. 1. Four face elements (a) face contour, (b) nose, (c) mouth, and (d) eye.

where

$$H_f = 1{\cdot}2P_5^{2{\cdot}59} + 1{\cdot}8 \tag{4}$$

The coordinate points (X_{f2}, Y_{f3}) and (X_{f2}, Y_{f4}) that are defined by Eqn. (5) are used to draw the hair boundary line.

$$\begin{aligned} X_{f2}/F_s &= 1 - u^{H_f} \\ Y_{f3}/F_s &= (1 - 0{\cdot}5P_6)Y_{f2}/F_s + 0{\cdot}125P_6\{4(1-u^{H_f})^3 + 6(1-u^{H_f})^2 \\ &\quad + (1-u^{H_f}) - 1\} \quad 0 < X_{f2} < 0{\cdot}5F_s \\ Y_{f4}/F_s &= (1 - 0{\cdot}5P_6)Y_{f2}/F_s + 1{\cdot}5P_6\{1 - (1-u^{H_f})\}^3 \\ &\quad 0{\cdot}5F_s < X_{f2} < F_s \end{aligned} \tag{5}$$

The auxiliary variable u employed in Eqns (2) to (5) ranges from 0 to 1 to generate fine-line elements on the CRT when drawing the face.

Using the variables shown in Figs 1(b) to (d), the nose, mouth and eyes are drawn inside the face contour so that their configuration is similar to that of a human face. These variables will now be mathematically defined for the nose, mouth and eyes. For the nose, defining the angle θ_n as

$$\theta_n = \tan^{-1}(2P_1 - 1) \tag{6}$$

and a nose-reference point $N = (N_x, N_y)$ as

$$\begin{aligned} N_x &= F_x + 0{\cdot}6(P_1 - 0{\cdot}5)F_s \\ N_y &= F_y + 0{\cdot}125(P_2 - 0{\cdot}5)F_s - 0{\cdot}15F_l F_s - N_l F_l(1 - \cos\theta_n) \end{aligned} \tag{7}$$

where

$$\begin{aligned} N_s &= 0{\cdot}33\exp(0{\cdot}5P_8 - 0{\cdot}25)F_s/\{8(1 - P_7)^3 + 2\} \\ N_l &= 0{\cdot}33\exp(0{\cdot}5P_8 - 0{\cdot}25)F_s - N_s \end{aligned} \tag{8}$$

the location of points N_1, N_2, N_3, and N_4 can be specified by the coordinates (X_{ni}, Y_{ni}), $(i = 1$ to $4)$

$$N_1: \quad \begin{aligned} X_{n1} &= \operatorname{sign}\theta_n \cdot N_s \cos 2\theta_n/\cos\theta_n \\ Y_{n1} &= F_l N_s \sin 2|\theta_n|/\cos\theta_n \end{aligned} \tag{9}$$

$$N_2: \quad \begin{aligned} X_{n2} &= N_l \sin\theta_n + \operatorname{sign}\theta_n \cdot N_s \cos\theta_n \\ Y_{n2} &= F_l(N_s \sin|\theta_n| - N_l \cos\theta_n) \end{aligned} \tag{10}$$

N_3: to be determined by the relationships
$N_3N_4 \perp N_1N_4$ and $N_sN_3 \perp N_3N_4$ (11)

$$N_4: \quad \begin{aligned} X_{n4} &= -\operatorname{sign}\theta_n \cdot (N_s + N_l \sin|\theta_n| \cos 2\theta_n) \\ Y_{n4} &= -(N_s + N_l \sin|\theta_n|) \sin 2|\theta_n| \cdot F_l \end{aligned} \tag{12}$$

The points N_1 and N_2, and N_3 and N_4 are connected by straight-line elements, and the points N_2 and N_4 by a circular line defined as

$$\begin{aligned} X_n &= \operatorname{sign}\theta_n \cdot N_s \sin\varphi_n \\ Y_n &= N_s \cos\varphi_n \cdot F_l \end{aligned} \tag{13}$$

where φ_n is an auxilliary variable ranging from $|\theta_n|$ to $2|\theta_n| - \pi$.

A reference point for mouth $M = (M_x, M_y)$ is defined as

$$\begin{aligned} M_x &= F_x + 0{\cdot}5(P_1 - 0{\cdot}5)F_s \\ M_y &= F_y - 0{\cdot}85F_l F_s + 0{\cdot}125(P_2 - 0{\cdot}5) \end{aligned} \tag{14}$$

for the given values (F_x, F_y) and F_s. Introducing M_g and M_f that are defined as

$$M_g = P_{12} - 0{\cdot}5 \qquad M_f = 2P_{10} - 1 \tag{15}$$

the contour of the mouth or lip line can be drawn using the mathematical formula specifying the upper and lower lips:

$$\begin{aligned} X_m/F_s &= 0{\cdot}3\exp(P_9 - 0{\cdot}5)u \\ Y_{mu}/F_s &= 0{\cdot}125\exp(P_9 - 0{\cdot}5)F_l\{M_f u^2 + M_g u + P_{11}(1-u^2) - M_c\} \\ Y_{ml}/F_l &= 0{\cdot}125\exp(P_9 - 0{\cdot}5)F_l\{M_f u^2 + M_g u - P_{11}(1-u^2) - M_c\} \end{aligned} \tag{16}$$

where M_c is determined by the values of P_{10} and P_{11},

$$M_c = \begin{cases} 0 & P_{11} \geq |2P_{10} - 1| \\ 0{\cdot}5(P_{11} - |M_f|) & M_f \leq 0 \\ 0{\cdot}5(|M_f| - P_{11}) & M_f > 0 \end{cases} \tag{17}$$

For the given values of F_x, F_y and F_s, the right eye and eyebrow can then be drawn using the parameters P_1, P_2, P_{13}, P_{25} and P_i (i = even number from 14 to 30). The x and y coordinates of the center point E between the right and left eyes are defined as

$$\begin{aligned} E_x &= F_x + 0{\cdot}4(P_1 - 0{\cdot}5)F_s \\ E_y &= F_y + 0{\cdot}2(P_2 - 0{\cdot}5)F_s \end{aligned} \tag{18}$$

Using the coordinates E_x and E_y, each center is then determined as

$$\begin{aligned} E_{xl} &= E_x - 0{\cdot}1(P_{13} + 4)F_s \\ E_{yl} &= E_y + 0{\cdot}2(P_{27} - 0{\cdot}5)F_s \end{aligned} \tag{19a}$$

$$\begin{aligned} E_{xr} &= E_x + 0{\cdot}1(P_{13} + 4)F_s \\ E_{yr} &= E_y + 0{\cdot}2(P_{25} - 0{\cdot}5)F_s \end{aligned} \tag{19b}$$

Introducing the inclination angle θ_e of the right eye and two parameters E_w and E_f,

$$\theta_e = P_{20} - 0{\cdot}5(\text{rad}) \tag{20}$$

$$\begin{aligned} E_w &= 0{\cdot}3P_{14} + 0{\cdot}85 \\ e_f &= 2P_{18} - 1 \end{aligned} \tag{21}$$

the eye shape can be drawn by Eqn. (22)

$$\begin{aligned} Z_e/F_s &= 0{\cdot}2uE_w \\ W_{eu}/F_s &= 0{\cdot}125E_w\{E_f u^2 + (1-u^2)P_{16} - E_c\} \end{aligned} \tag{22}$$

where u is the auxiliary variable ranging from 0 to 1, and E_c is defined as

$$E_c = \begin{cases} 0 & P_{16} \geq |2P_{18} - 1| \\ 0{\cdot}5(P_{16} - |E_f|) & E_f \leq 0 \\ 0{\cdot}5(|E_f| - P_{16}) & E_f > 0 \end{cases} \tag{23}$$

The center point of the right pupil is determined by the coordinates A_{xr} and A_{yr}

$$\begin{aligned} A_{xr} &= E_{xr} + 0{\cdot}1E_w(2P_{22} - 1)F_s \\ A_{yr} &= 0{\cdot}075E_w(2P_{24} - 1)F_s + E_y \end{aligned} \tag{24}$$

The distance between the eye and eyebrow is determined by the number B_h defined as

$$B_h = 0{\cdot}075(2P_{28} + 1)F_s \tag{25}$$

and thc right eyebrow is drawn using two variables Z_b and W_b

$$\begin{aligned} Z_b &= 0{\cdot}2uF_s \\ W_b &= 0{\cdot}025(1+u)[1 + 2B_f(u-1) - \{0{\cdot}5(1+u)\}^7]F_s \end{aligned} \tag{26}$$

where

$$B_f = 3P_{30} - 1 \tag{27}$$

The left eye and eyebrow are drawn in a similar way, but the parameters P_i (i = even number from 14 to 30) are replaced with P_j (j = odd number from 15 to 31).

Face-Graph Coloring

The microcomputer we used has eight basic colors—blue, red, violet, green, light blue, yellow, white and black—40 other intermediate colors were produced by specifying different colored-dot combinations. These 48 colors were analyzed to clarify their hue, value and chroma, and a colored body using each of the 48 colors generated by the microcomputer was constructed to evaluate the color gradation.

The eye pupils, brows and hair are painted in black; the mouth red; and the skin in such a color as blue, pink, or red, which is specified by

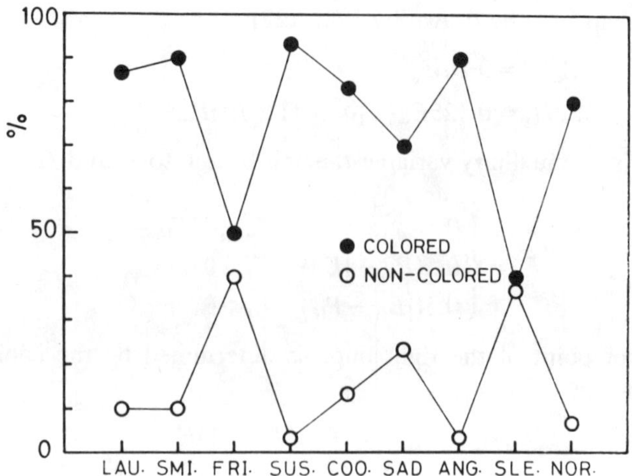

FIG. 2. Color effect reinforcing facial expressions.

a parameter value representing the global state of the system being monitored. Great care must be paid in selecting the representing coordinate point for each facial element, which is always included in the element to paint the face properly. Plate 1a is a comparison between the monochrome and colored face graphs thus generated.

The monochrome face graphs express nine states of the human psyche, i.e. laughter, smiling fright, suspicion, coolness, sadness, anger, tiredness, and 'normalcy'. [1, 2] For these facial expressions, we examined whether color could be used to intensify these expressions. A visual comparison between the monochrome and colored face graphs drawn on a color CRT gave the results in Fig. 2, in which seven colored graphs—excluding fright and sleepiness—were perceived as being more intense than their noncolored representations. Thus, we concluded that adding the proper color to the face can reinforce the facial expression, and we took this color effect into account when displaying plant safety conditions.

DISPLAY OF SAFETY CONDITIONS [3]

A demonstration of colored face graphs in displaying the safety conditions of a large industrial system was done for the case of the

---

PLATE 1. (a) Non-colored and colored face graphs. (b) Integrated colored face graph display of the TMI accident.

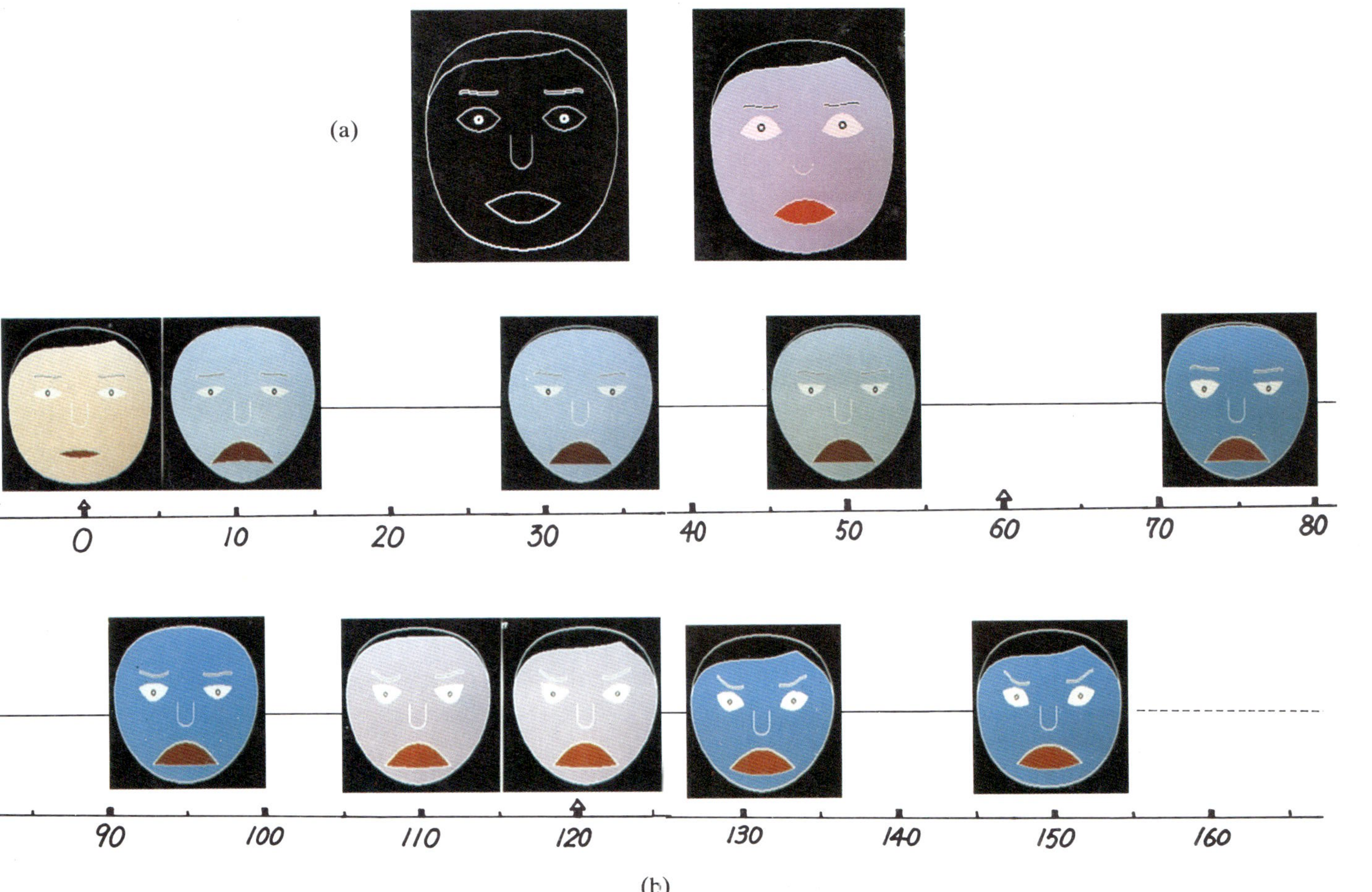
(a)
0
10
20
30
40
50
60
70
80
90
100
110
120
130
140
150
160
(b)

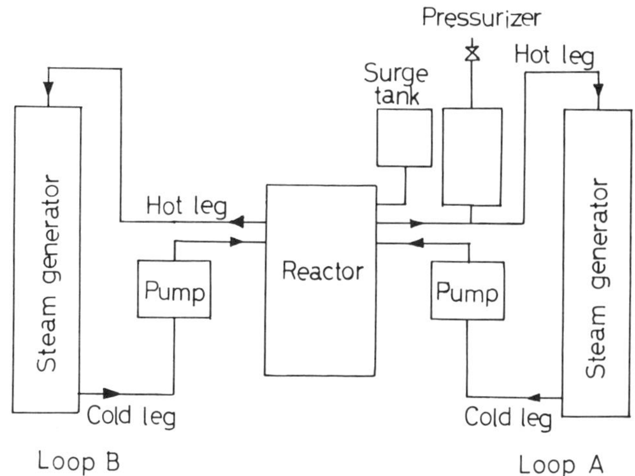

FIG. 3. Schematic flow diagram of the TMI No. 2 reactor plant.

Three Mile Island (TMI) nuclear power plant accident, and its characteristics were examined with respect to effects of a colored face graph on human visual perception.

Background of the TMI Accident

According to reports [4] a loss of feedwater to the TMI facility's steam generator at about 4 a.m. on 28 March, 1979, resulted in a turbine trip. Shortly after the reactor protection functioned as designed, the steam generator water dropped to an emergency level, but was not resupplied due to a valve closed by mistake. A pilot-operated relief valve remained open and undetected for about 2 h, contributing to the continued loss of coolant. Figure 3 schematically illustrates TMI's No. 2 reactor plant, whose major components were a pressurized water reactor, steam generator, pressurizer and reactor coolant pump. The plant had two coolant circulation loops, A and B.

Fifteen kinds of thermohydraulic data were recorded during the accident; the coolant temperatures in the hot and cold legs, the water level in the pressurizer, and the pressure of the main coolant are given in Figs 4(a) and (b). The temperature in the hot legs exceeded 620 °F (327 °C) because the main pump had been turned off due to its severe vibration. The flow rate of the main coolant in both loops decreased moderately preceding this, but apparently hit zero the instant the pump was turned off.

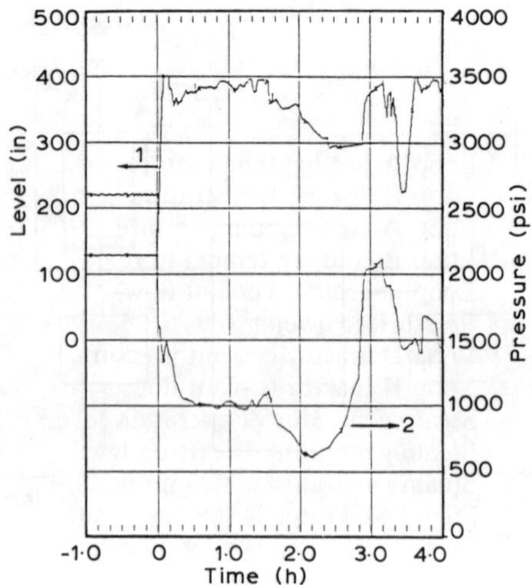

FIG. 4(a). Records of the pressurizer level and coolant pressure during the TMI accident.

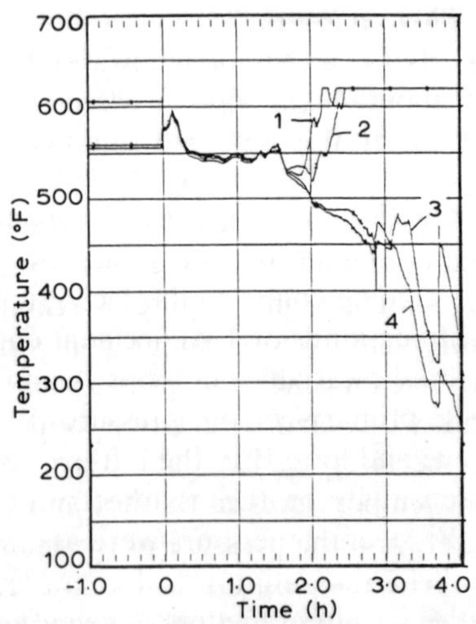

FIG. 4(b). Time records of temperature in the hot and cold legs of the TMI plant.

TABLE 2
Fifteen kinds of TMI data recorded

Channel no.	*Plant variables*
1	Loop A hot leg temperature
2	Loop B hot leg temperature
3	Loop A cold leg temperature
4	Loop B cold leg temperature
5	Loop A reactor coolant flow
6	Pressurizer level
7	Loop B reactor coolant pressure
8	Loop B reactor coolant flow
9	Steam generator A operation level
10	Steam generator A start-up level
11	Steam generator A steam pressure
12	Steam generator B steam pressure
13	Steam generator B operation level
14	Steam generator B start-up level
15	Composite primary coolant pressure

Assignment of Plant Data to Colored Face Graph Parameters

Our colored face graph has 31 parameters $P(i)$: = 1 to 6 for the face contour with hair, 7 and 8 for the nose, 9 to 12 for the mouth, and 13 to 31 for the eyes, taking a value between 0 and 1. The face displays different parameter states in the form of facial expressions. One parameter is used to determine the face color, which must be defined as describing the global state of the plant, taking into account the 31 parameter values. Note that care is needed to make the facial expression consistent with human feelings perceived from the colors; we used the results of studies on the relationship of psychological states and colors in assigning colors to the face graph.

Table 2 gives the fifteen kinds of TMI accident data, which consisted mainly of thermohydraulic variables in loops A and B, the pressurizer level, and a composite primary-coolant pressure. We assigned data for loop A to the right eye and loop B to the left eye to take advantage of man's keen visual sensitivity to the asymmetry of facial expression. The water levels in the steam generators were assigned to the hair and chin, which provide highly sensitive visual cues [5]. The result of the data assignment to the facial parameters is given in Table 3; any face parameters undescribed were assigned a constant value of 0·5.

Original data on the TMI accident covered different value ranges,

TABLE 3
Assignment of the 15 data to fit face graph parameters

Channel no.	*Face graph parameters P(i)*
1	Right eye shape 18
2	Left eye shape 19
3	Inclination in the right eye 20
4	Inclination in the left eye 21
5	Right eye openness 16
6	Mouth shape 10
7	Mouth size 9
8	Left eye openness 17
9	—
10	Amount of hair 6
11	Shape of the right brow 30
12	Shape of the left brow 31
13	—
14	Jaw shape 4
15	Mouth openness 11

depending on the plant variables. Therefore, we have normalized the data so that it takes a value of 0·5 for a variable's normal state, the highest and lowest data values being 1 and 0. This data assignment was used to generate face graphs to display plant safety states by facial expressions differing from the normal. All the face parameter values used thus took values close to 0·5 when the plant was running very well, and near 0 or 1 for a dangerous plant state.

The departure of a face parameter value from the normal 0·5 seems to be a global indicator of the plant safety condition. Thus, we used the average, *a*, of the 15 absolute vlaues $Q(i)$ defined by the face parameters as

$$\begin{aligned} Q(i) &= 1 - |(P(i) - 0{\cdot}5) \times 2| \qquad i = 1, 2, \ldots, 15, \\ a &= (Q(1) + Q(2) \cdots + Q(15))/15 \end{aligned} \tag{28}$$

The value of *a* is 1 when the plant runs normally, and approaches 0 under catastrophic plant conditions. Dividing the range of the value *a* into tenths, we assigned the following colors to each subrange of the value of *a*, i.e.,

$$a = 1 \cdots 0{\cdot}8 \cdots 0{\cdot}6 \cdots 0{\cdot}4 \cdots 0{\cdot}2 \cdots 0$$

SC LV PG B V

where the symbols stand for the following colors: SC, skin color; LV, light violet; PG, purple grey; B, blue; and V, violet.

Note that the colors were selected from the color body to obtain a color gradation (A) changing hue gradually from a skin color to violet, and the face graph was designed to express a hopeless expression when the plant entered a catastrophic state—a face with little hair, a narrow chin, a turned-down mouth, pulled-togehter eyebrows, and a violet color.

Results

Digitized values from the 15 channels of the TMI No. 2 reactor plant were obtained from the records, [4] where the sampling time step was 5 minutes and data length $2\frac{1}{2}$ h. Plate 1(b) is the time series for an integrated color face graph display of the TMI accident, where the horizontal axis is the time number denoting 5-min intervals and each arrow is the range in which the same face graph appears.

At time instant 3, the jaw becomes small, the hair a little less than normal, and the color slightly violet, indicating that the face is bothered, and in trouble. At the time instant 4, the face looks very embarrassed or in serious trouble, corresponding to the reactor condition and the greatly reduced pressurizer water level. At time instant 10, a deep anger appears in the eyes of the pale face, indicating that the temperature of the hot leg coolant was gradually increasing. At time instant 19, the face changes suddenly to blue, expressing fright corresponding to the abrupt decrease in the loop A coolant flow rate. For time instants 24 to 29, the face expresses severe stubborn endurance at full strength, but still indicating fear in the eyebrows. This is due to the high temperature and very low flow rate in the hot legs. From the time instants 29 to 33, the face still shows fright, but with tentative positiveness as the water level recovered in the steam generators.

EVALUATION

Three aspects were used to analyze the human visual perceptive characteristics of the color face graphic display as applied to the TMI accident: (1) how well the plant can be perceived from the face graph, (2) how natural the change of color in the series of displays is felt to be, and (3) how consistent the face configuration and color are felt to

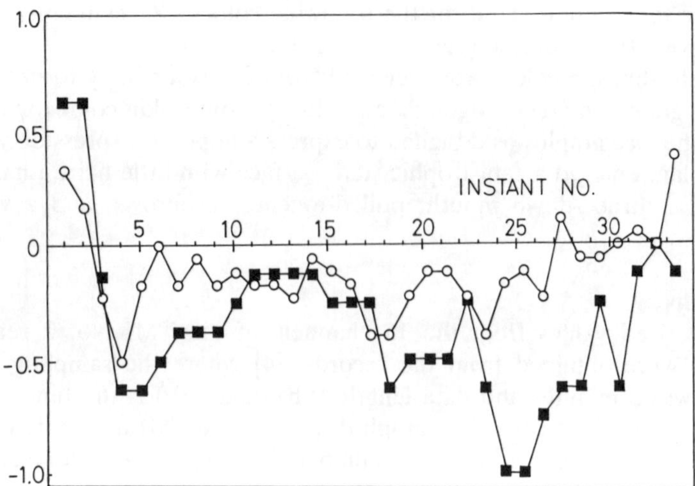

FIG. 5. Human perception to the TMI accident process, the count 1 being very good and −1 very bad (solid symbol, colored display; open one, monochrome display).

be. For each aspect, the best score was assigned a count of 2, a fair score a count of 0, and the worst score −2. Note that we also tested aspect 1 without changing the face color, i.e., the face was an ordinary skin color like the first one in Plate 1b and the face graphs were compared with and without any color change.

Ten male students were employed in the test. We averaged the total scores for each time instant and normalized these values at 1 for the best and −1 for the worst.

Figure 5 is the result of human perception of the TMI accident. Solid squares are for a colored face graph display with color change, and open circles for those without a color change. In the first stage when the plant fell into the reactor scram, the change in plant condition was more vividly perceived without a color change. For the plant in the worst condition during time instants 24 to 30, the colored face graph was able to emphasize the plant condition more strongly than graphs without a color change.

For aspects (2) and (3), another color gradation (*B*), starting from skin color and moving to pink, yellowish green, yellow, red, and violet, was also tested to investigate the effect of color gradation on human visual perception. This color gradation was selected from the CRT's color body and changed hue gradually.

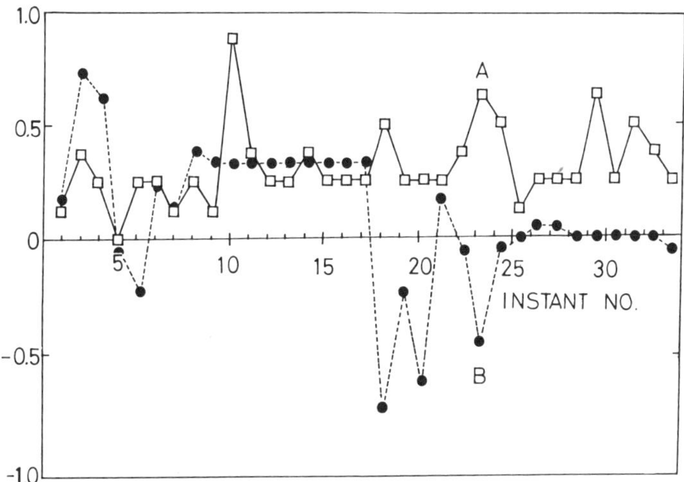

FIG. 6. Naturalness of color change, the count 1 being very good and −1 very bad (open symbol, gradation *A*; solid one, gradation *B*).

Figure 6 shows the results of human reaction to the change in face color during display of the accident process, where open squares are for the color gradation *A* and solid circles for the gradation *B*. The colors selected from gradation *A* were better perceived by human visual acuity than those from gradation *B*, even though the colors were less than realistic. This means that the human reaction to color is important in the design of colored face graphs, and may not need to be realistic to communicate the global condition of a plant to its operators.

Figure 7 is the result of human perception of the consistency between facial expression and color, where open circles are for color gradation *A* and solid squares for gradation *B*. The results from gradation *B* are generally poor, but those from gradation *A* are fairly good, even though the count is rather low during time instants 7 to 18, during which the face colors were pink, light violet and purple grey, which are not strongly related to the feeling of Japanese people, perhaps. Figure 7 indicates that emotional feeling perceived from colors is important in correctly transferring plant safety states to operators and in avoiding unnecessary misunderstanding of plant conditions.

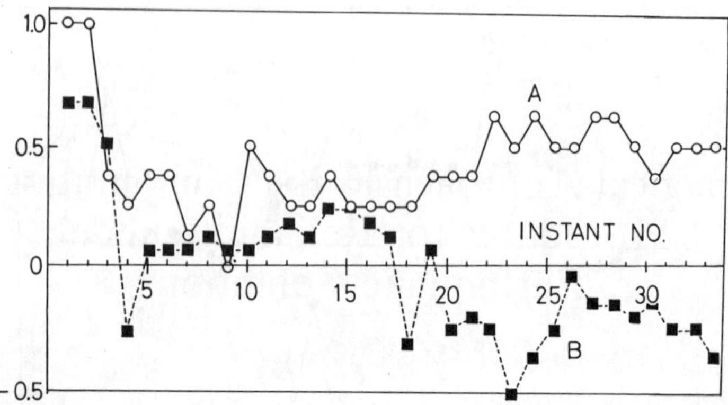

Fig. 7. Consistency between facial expression and color, the count 1 being very good and −1 very bad (open symbol, gradation *A*, solid one, gradation *B*).

CONCLUSION

This paper has examined human visual perception of colored face graph displays of the safety condition in a large, industrial system by employing the TMI plant accident process, and has demonstrated the usefulness of this integrated display in transferring the global safety condition of a complicated plant system during an accident process. Color was effective in emphasizing the plant system data when visually perceived by human beings. The colors used in an integrated colored face graph display must be completely consistent with the emotional feelings perceived from the colors, meaning that the colors need not necessarily be realistic.

REFERENCES

[1] F. Hara et al., Basic Study of Face Graph for Application to Plant Surveillance. *Trans. SICE,* **15**(1) 53 (1979).

[2] T. Yamashita and F. Hara, The Use of Face Graphs in Man–Machine Communications. *Proc. ICCS,* **2**, 1284 (1978).

[3] F. Hara, Integrated Color Face Graphs for Plant Accident Display, *2nd IFSA Congress Preprints,* Vol. 2, Tokyo, 807, July 1987.

[4] Staff Reports to the President's Commission on the Accident at Three Mile Island, **1,** 1 (1979).

[5] T. Yamashita and F. Hara, Dynamic Characteristics of a Face Graph for Displaying Plant Conditions, *SICE,* **18**(5) 525 (1982).

Practical Approach to the Computational Procedures for Reliability under Earthquake Conditions

MASARU ZAKO

Department of Mechanical Engineering and Technology Education, Mie University, Tsu city, Mie, Japan

ABSTRACT

A large number of studies on structural reliability have been made [1]. The most important omission, however, is that there are very few reports about practical problems.

The purpose of this paper is to show how to calculate the failure probability for practical applications. A 14 000 kl oil storage tank is given as the example and the failure probability of the tank when an earthquake has occurred is calculated. The earthquake level is estimated according to the design life and site of the tank, and an artificial earthquake wave based on the estimated earthquake level is produced by computer. The sloshing phenomenon of oil in the tank is analyzed, this sloshing acting on the time response of pressure on the tank wall. The time response of the stress and strain at each point in the tank is calculated by FEM, and it becomes clear that the annular plate is the area of maximum stress and strain loading.

As a result, the D-value as a failure probability is calculated from the time response of the strain, based on Miner's rule. It will be seen that the computational procedure described in this paper is very practicable.

INTRODUCTION

In order to obtain structural reliability and risk assessments under earthquake conditions, analyses of the failure of a structure and the risk of secondary failure are necessary. Especially, the optimum selection of the repair period and its cost can be calculated by risk analysis. The value for practical usage of such calculation procedures is well recognized and the number of related reports is increasing. However, there are very few reports about practical examples of structural reliability. In this paper, an oil storage tank is selected as an example of a structure, and the computational method to obtain the failure probability is discussed.

COMPUTATIONAL PROCEDURE

The flow chart for the computational procedure to assess the structural reliability is shown in Fig. 1. The details of each routine in Fig. 1 will be described in the following sections.

Site of the Structure

An earthquake produces different loading characteristics at different locations, so it is necessary to determine the site of the structure in order to define the seismic load.

Estimation of the Seismic Load

If the site of structure has been determined, the expected value of the seismic load at the site can be calculated. The calculation methods will not be discussed here since there exist many books which deal with these topics in detail [2, 3].

Modeling the Structure

A model of the structure is necessary for an analysis of stress and strain by using FEM, and it is important to assume the fragility sequence to analyze the whole structural model. However, it will take much computation time if the whole structural model is used for stress–strain analysis, therefore, it is practical to analyze only detailed areas which are defined by the fragility sequence.

Analyses of Stress and Strain

There are two methods for this analysis, one being a time history response analysis and the other a modal analysis. In respect of computational time, the modal analysis is more valuable than the other method. With this method, however, designers have to consider how many modes should be analysed, etc. In this paper, the time history response analysis has been used. A number of reports reveal that the Stochastic Finite Element Method [4] or NASTRAN are very useful for this purpose.

Comparison of Strength

A judgement of the safety margin is made by comparing the strength with the computational stress (or strain). If the structure that has been subjected to an earthquake is now in use, however, the cumulative damage during the service period needs to be calculated before adding the earthquake damage to it.

Cumulative Damage

The cumulative damage for a cyclic loading condition like an earthquake can be obtained from the values of stress or strain amplitude. As the total number of cycles of an earthquake wave which can damage a structure is less than 10^3 and the load level is large, the fatigue condition is in the low-cycle region and the strain criterion is more reasonable for computing the damage than the stress criterion. The cumulative damage rate is calculated by the relationship between the strain amplitude and number of cycles to failure according to Miner's rule.

Failure Probability

The distribution and scatter of strength have to be elucidated in order to calculate the failure probability. On the other hand, the strains at each point in the structure that have been obtained by the Finite Element Method are determinate values without scatter. Therefore, the failure probability of the structure can be calculated only by the distribution of material strength. The expected values and standard deviations of strains can be calculated if the Stochastic Finite Element Method is used for strain analysis under the applied load.

Consequently the failure probability can be obtained by the distributions of strength and strain.

COMPUTATIONAL RESULTS

In order to apply this foregoing computational procedure, a 14 000 kl oil storage tank was chosen. With the flow chart in Fig. 1, the procedure can be developed as follows.

Site and Estimation of Earthquake Level

Tokyo is selected as the construction site of the 14 000 kl oil storage tank. From the view points of structural life and reasonable manufacturing cost of the tank, the return period of earthquakes which will

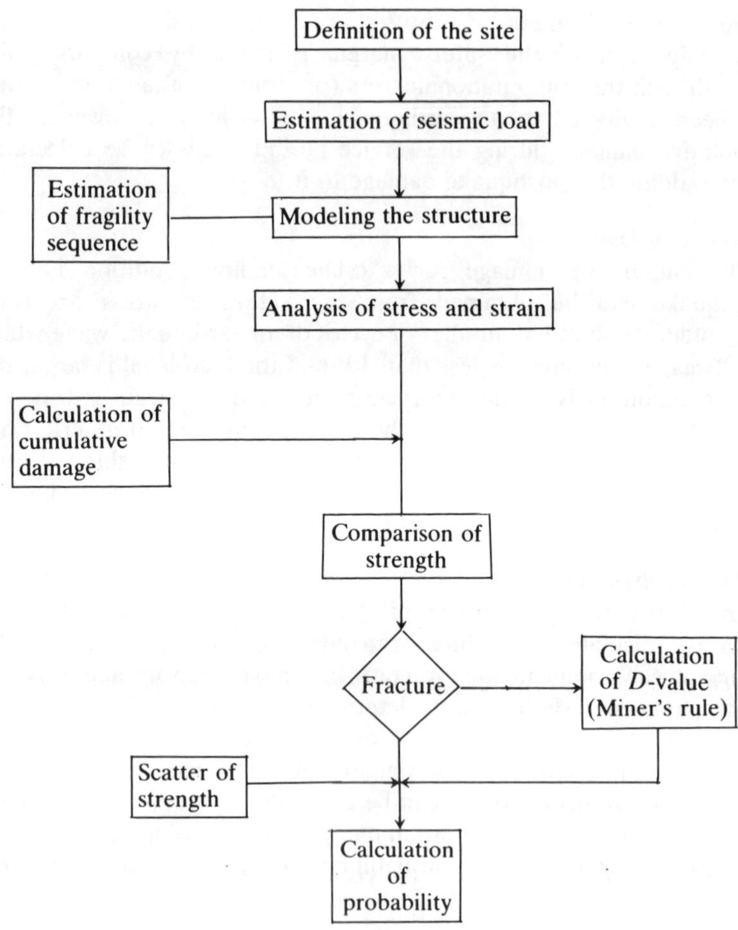

FIG. 1. Flow chart for the computational procedure.

attack the tank is assumed to be 50 years. The power spectrum of an earthquake can be obtained by the site and the return period. Figure 2 shows the power spectrum of an earthquake with a return period of 50 years in Tokyo. The artificial earthquake wave can be created from this power spectrum by using a computer program developed for the purpose. The envelope curve to the earthquake wave has to be determined beforehand, and that adopted in this paper is shown in Fig. 3; this is the most common earthquake wave selected in Japan. The artificial earthquake waves obtained are presented in Figs. 4–6, in

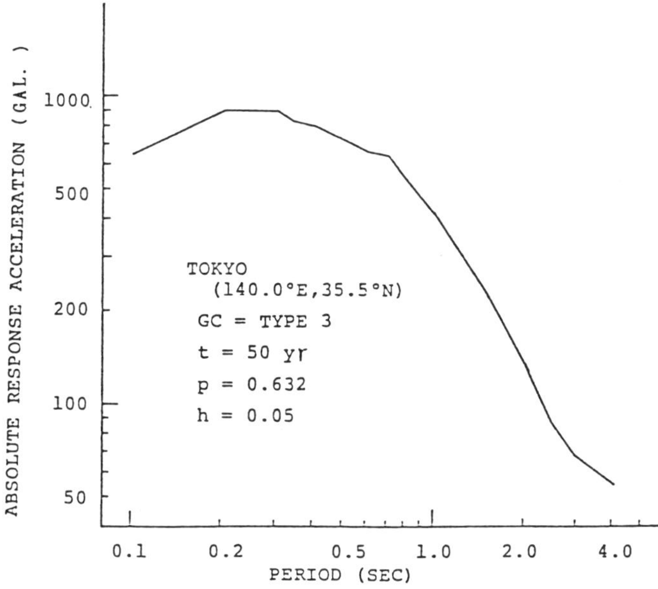

FIG. 2. Acceleration response spectrum.

which the time histories of acceleration, velocity and displacement are shown, respectively.

Structure Modeling

Failure is generally caused by the seismic sloshing of oil, so that stress and strain in the annular plate of the tank, where the highest values will occur, have to be analyzed. Therefore, the following two

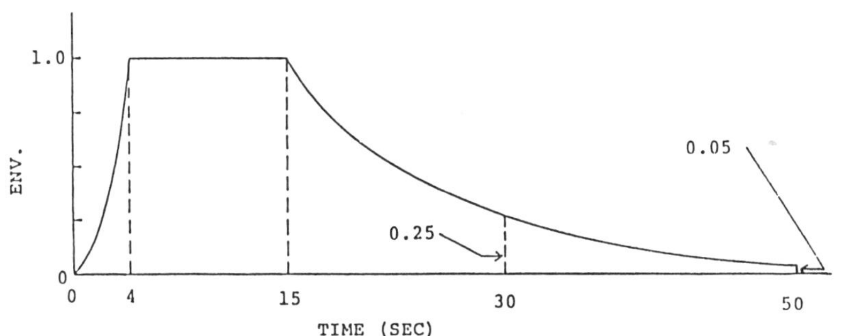

FIG. 3. Earthquake envelope.

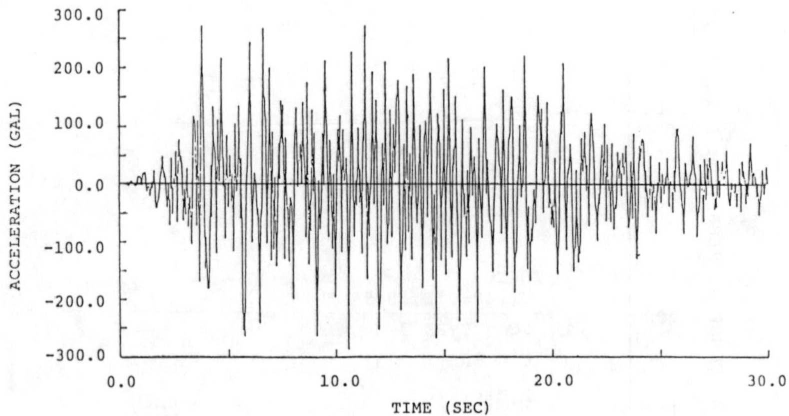

FIG. 4. Artificial earthquake wave (acceleration).

models should be considered.

(1) Model for sloshing analysis.
(2) Model for stress and strain analysis of the annular plate of the tank.

The pressure conditions at two instants under earthquake loading are shown in Figs. 7 and 8. The time history of the pressure acting on the tank wall, which is applied to the annular plate as an external

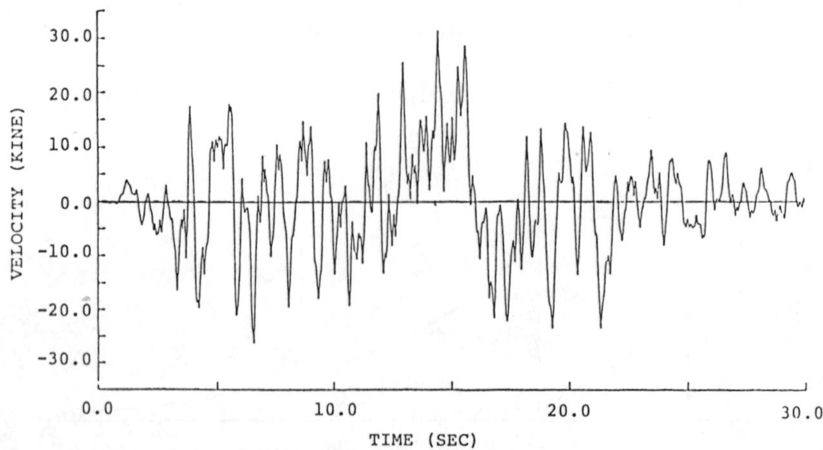

FIG. 5. Artificial earthquake wave (velocity).

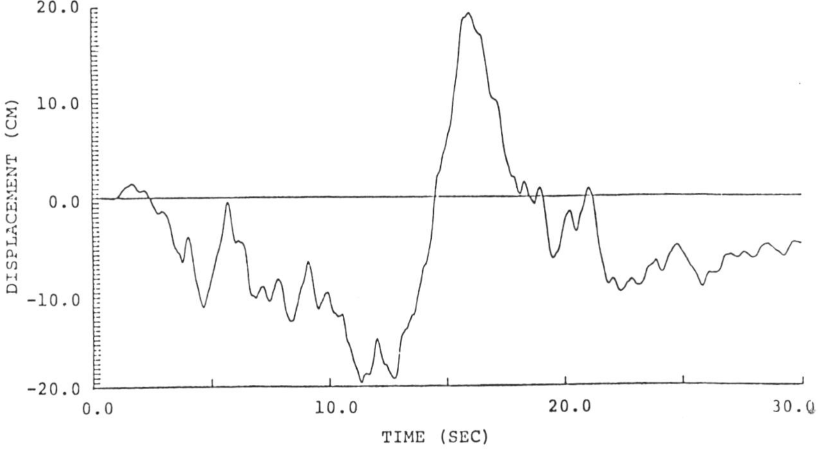

FIG. 6. Artificial earthquake wave (displacement).

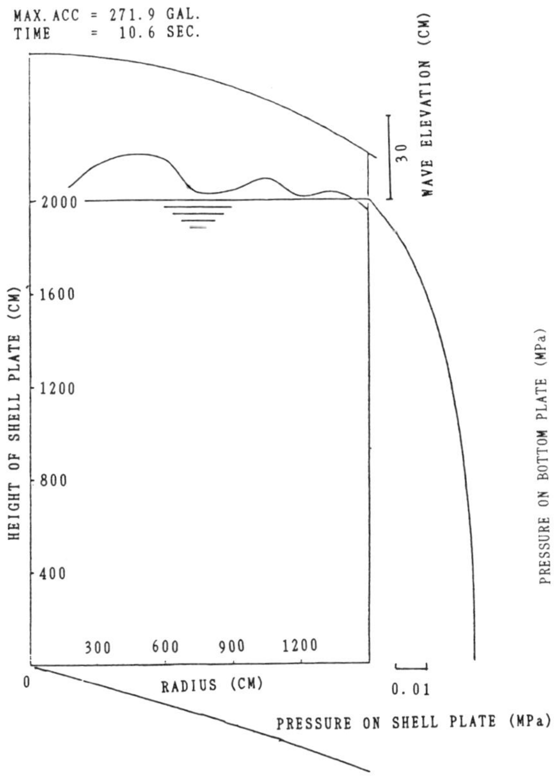

FIG. 7. Example of sloshing at 10·6 s.

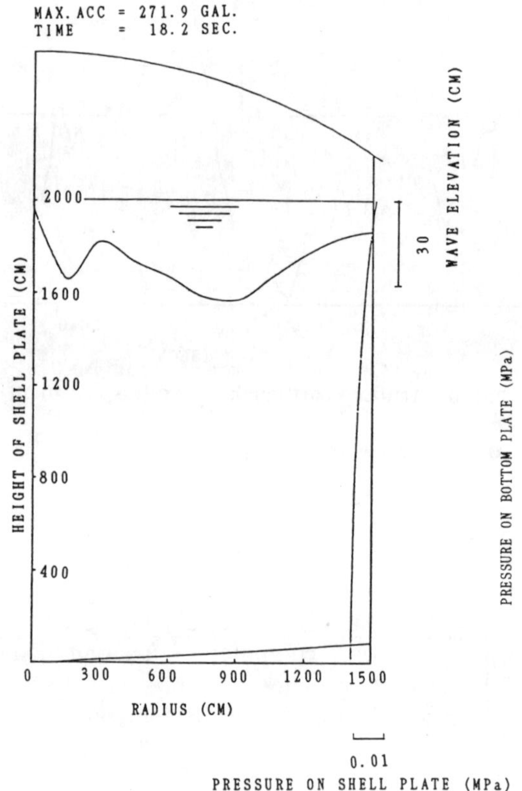

FIG. 8. Example of sloshing at 18·22 s.

force, is shown in Fig. 9. The stress and strain on the annular plate can be calculated by using the Finite Element Method for the FEM model shown in Fig. 10. The time histories of stress and strain are calculated with the pressure acting on the node points of the FEM model.

Time History Response Analysis

In this paper, time history responses were analyzed by the Stochastic Finite Element Method. The calculated values of stress at every 0·5 s under earthquake loading are shown in Table 1. Because the stress level is low enough for the relationship between stress and strain to be linear, the cumulative damage by Miner's rule can be calculated by using either the stress or strain. When the load level becomes higher, the cumulative damage should be calculated by strain.

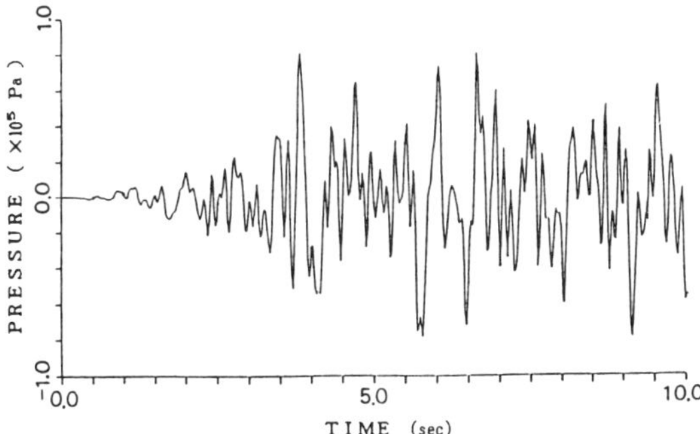

FIG. 9. Calculated pressure response.

Calculation of the Cumulative Damage

Miner's rule defines cumulative damage D by

$$D = \sum_{i=1}^{n} \frac{n_i}{N_i} \tag{1}$$

Failure occurs under the condition of $D = 1{\cdot}0$. The D-value can be calculated using a personal computer or a larger computer. In this paper, the D-value was calculated by using the values in Table 1, the computational results being shown in Fig. 11. The damage rate increases rapidly at the early stage of the earthquake and then increases slightly after 10 s. It has been assumed that the calculated results of stress belong to a normal distribution. The expected value and deviation of the stress have been obtained, so that the D-values of $\pm 3\sigma$ (standard deviation) can then be obtained. Figure 11 shows the relationship between the D-values of $\pm 3\sigma$ and time. As the failure occurs at $D = 1{\cdot}0$, it can be recognized that the result of $D = 10^{-4}$ to 10^{-6} which was obtained is sufficiently safe.

CONCLUSIONS

The computational procedure for structural reliability under earthquake loading has been explained by using an example. There are many computer programs which can calculate the time history response of stress and strain, including NASTRAN and MARC. However, the cumulative damage rate cannot be calculated by these

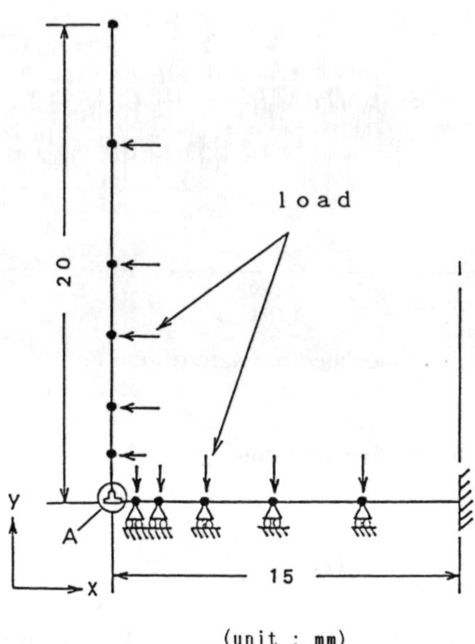

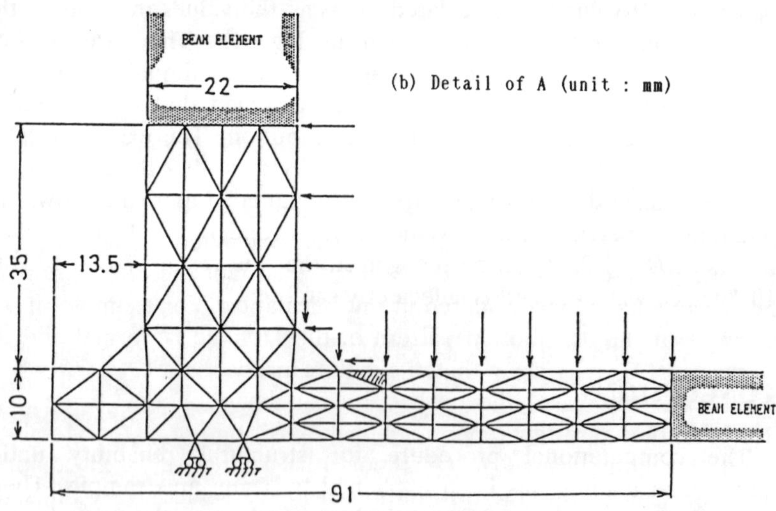

FIG. 10. FEM model.

TABLE 1
Expected value and variance of stress

Time (s)	*Expected value (MPa)*	*Variance $(MPa)^2$*
0·5	−0·539	$4{\cdot}88 \times 10^{-4}$
1·0	4·24	$3{\cdot}03 \times 10^{-2}$
1·5	2·09	$7{\cdot}26 \times 10^{-3}$
2·0	25·5	1·09
2·5	−6·06	$6{\cdot}13 \times 10^{-2}$
3·0	−5·25	$4{\cdot}60 \times 10^{-2}$
3·5	50·3	4·26
4·0	−50·7	4·31
4·5	−23·5	$9{\cdot}23 \times 10^{-1}$
5·0	−8·67	$1{\cdot}28 \times 10^{-1}$
5·5	61·9	6·43
6·0	92·5	$1{\cdot}44 \times 10$
6·5	−93·2	$1{\cdot}46 \times 10$
7·0	−70·8	8·41
7·5	41·3	2·87
8·0	−69·6	8·13
8·5	74·2	9·28
9·0	24·0	$9{\cdot}70 \times 10^{-1}$
9·5	71·4	8·61
10·0	−100·0	$1{\cdot}63 \times 10$

programs, and a computer program which can calculate this is necessary. A computer program for calculating the D-value by Miner's rule and the S–N curve was written for reporting in this paper. If the material under fatigue loading will fail when $D = 1{\cdot}0$, the probability of failure of the material can be easily calculated. However, the strength of materials has a wide dispersion, so that the distribution of the D-value must be considered in the calculation. For example, if we assume that the distribution pattern of the D-value is normal and its average is 1·0, it is clear that the failure probability P_f is 0·5 when $D = 1{\cdot}0$. In this case, the probability can be obtained very easily. However, there are few reports about the distribution of the D-value, and the average value of D is not 1·0. Therefore, a calculation procedure considering the distribution of the D-value is very difficult. As a result, it can be recognized that the method in which the structural reliability is explained only by the D-value as described in this paper is satisfactory in practice.

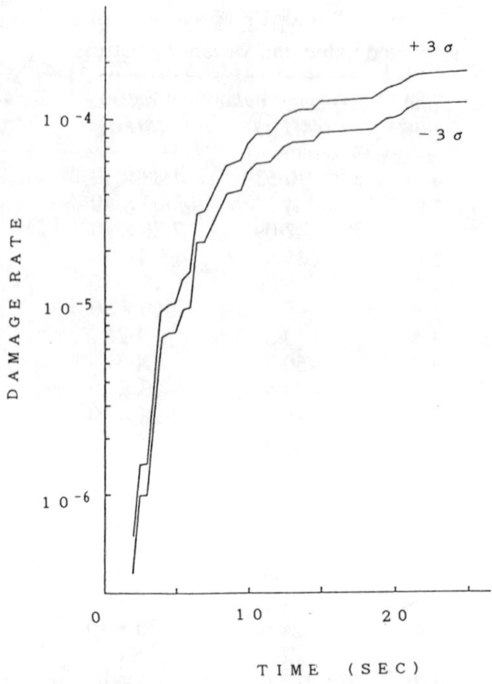

FIG. 11. Variation of the *D*-value with time.

ACKNOWLEDGEMENTS

The author wishes to express his gratitude to Dr. Yoshio Ochi for helpful discussions, and to Dr. Shigeru Shimosaka and Dr. Nobuyuki Kobayashi, all of the Research Institute of Ishikawajima-Harima Heavy Industry Co. Ltd., who contributed computational efforts for an artificial earthquake. The author also would like to acknowledge the considerable assistance of Professor Toshiro Miyoshi of Tokyo University.

REFERENCES

[1] Y. J. Park, A. H.-S. Ang and T. K. Wen, Seismic damage analysis and damage-limiting design, *Structural Safety and Reliability*, Vol. II, IASSAR, New York, 197 (1985).

[2] J. Sakamoto, T. Kohama and Y. Mori, A statistical representation of earthquake load effects of use in reliability-based design, *Structural Safety and Reliability*, Vol. II, IASSAR, New York, 217 (1985).
[3] T. Ohta and H. Ando, Averaged ground motion for the site condition of ground, *Proceedings of the 5th Symposium on Ground Vibrations*, Architectural Institute of Japan, 27 (1977).
[4] T. Hisada and S. Nakagiri, Role of stochastic finite element method in structural safety and reliability, *Structural Safety and Reliability* Vol. I, IASSAR, New York, 385 (1985).

Index

Alumina ceramics
bending tests, 73, 78
chemical composition, 72–4
cumulative failure probability, 74–82
flexural fracture, 74–82
fracture surface, 78
fracture toughness, 81
impact tests, 73
strength against static and impact loads, 71–83

Bayesian reliability analysis, 167–212
analytical model and formulation, 169–72, 178–83
application to field data, 168–77
comparison with field data, 173
proper inspection interval based on, 178–85
Bending tests, alumina ceramics, 73, 78
Bienayme–Chebyshev inequality, 19
Bounding operations, 94
Branching operation, 92–3
Bridge structures
reliability analysis, 119–34
main girder, 122–4, 129–32

Ceramics
brittleness of, 71
strength variation, 71
see also Alumina ceramics

Color-face graph parameters, 207
Color-face graphic display, evaluation of, 209–11
Color-face graphic generation, 198–204
Combined-cycle power generating plant, 3
Composite variability model
comparison with experimental results, 16–18
parameter randomization, for, 12
prediction by, 13–16
Crack detection probability, 160, 179
Crack length distribution, 153
Crack propagation life, 143, 144
gas turbine components, 7
Cumulative damage, calculation of, 221

D-value, 221, 223
Damaged redundant structures
damage measurement, reliability analysis of, 106–9
failure probability, 109–13
reliability analysis, 105–17
safety index, 107
safety margin, 107
Defect detection probability, 10
DEST-I, 193

Earthquake conditions, reliability analysis, 213–25
Equivalent nodal forces, 86–90

Face-graph coloring, 203–4
Face-graph drawing, 198–203
Fail-safe crack length, 180
Failure mechanisms, portal frame, 113–14
Failure probability, *See* Probability of failure
Fatigue crack growth, 157
  reliability analysis, 153–64
Fatigue crack growth rate, 12–18
Fatigue crack length, 169, 180
Fatigue crack propagation, 179
  reliability analysis, 135–51
Fatigue life, 169, 171, 188
Fatigue-proof design, 25–50
Fiber-reinforced composite materials
  behaviour under combined stress states, 52–3
  design theory, 52–3
  maximum stress criterion, 54–5
  maximum work criterion, 55–7
  mechanical properties, 51
  probabilistic design, 51–70
  probability of failure, 54, 57, 58
  stress analysis of notched plate, 58–62, 67–9
Finite Element Method, 214, 220
Flaw size distribution
  initial, 8–10
  significance of, 7–8
Fleet reliability, 30–49
Fokker–Planck equation, 154, 155
Fracture model, 136
Fracture surface, alumina ceramics, 78
Fracture toughness, alumina ceramics, 81
Frame structures
  generation of safety margins, 90–2
  reliability assessment, 85–117
  structural failure criterion, 90–2
  structural failure modes under combined loads, 86–92
  ultimate collapse analysis, 85–117

Gas turbine components, crack propagation life, 7
Gas turbines, maintenance control, 3–7
Gumbel probability paper, 125

Hanshin (Osaka–Kobe) Expressway Public Corporation, 120
Heuristic operations, 94
Highway bridges, reliability analysis, 121–7

Impact tests, alumina ceramics, 73
Industrial system, safety conditions, 197–212
In-service inspection (ISI), 137, 150, 153, 157, 167–212
Integrated color graphics technique, 197–212

Japan Classification Society for Ships (NK), 97
Joint probability density function, 121–2

Knowledge engineering
  comparison with conventional approach, 192–4
  debugger or tracer of knowledge, as, 194–5
  reliability improvement, for, 191–212

Linear elastic fracture mechanics, 21

Maintenance and repair program, 105
Maintenance control, gas turbines, 3–7
Markov approximation method, 154
Markov chain, 161, 185
  reliability analysis of fatigue crack propagation, 135–51
MAST-I, 193

Maximum likelihood estimate (MLE), 27, 28, 36, 39
Maximum stress criterion, 54–5
Maximum work criterion, fiber-reinforced composite materials, 55–7
Miner's rule, 220, 221, 223
Monte Carlo simulation, 27, 29, 37

NASTRAN, 214
Non-destructive inspection (NDI), 10, 137, 150, 188
  reliability of fatigue crack propagation, 135–51
Notched plate, stress analysis of, 58–62, 67–9

Oil storage tank, reliability under earthquake conditions, 215–21

Parameter randomization, composite variability model for, 12
Paris–Erdogan crack growth model, 155
Paris–Erdogan equation, 141
Paris's law, 12
P_f–N relationship, 146–8
PNET method, 109, 116
Portal frame, failure mechanisms, 113–14
Pre-service inspection (PSI), 139–40, 150
Probabilistic fracture mechanics, 1–24
Probabilistically dominant failure paths, 92–4
Probability density function (PDF), 159, 161, 162, 171, 188
Probability distribution function, 123–4
Probability of events, 170–1
Probability of failure, 186, 215
  damaged redundant structures, 109–13

Probability of failure—*contd.*
  fiber-reinforced composite materials, 54, 57, 58
  specified, 126–7, 130, 132
  upper bound approach, 18–22
Probability of fracture, 136, 139, 145

Reduced stiffness matrixes, 86–90
Reliability analysis
  Bayesian. *See* Bayesian reliability analysis
  bridge structures, 119–34
    main girders, 122–4, 129–32
  damaged redundant structures, 105–17
    damage measurement, 106–9
  earthquake conditions, 213–25
  fatigue crack growth, 153–64
  frame structures, 85–117
  highway bridges, 121–7
  space domain, 123
  time domain, 123–4
Reliability-based scatter factors, 25–50
Reliability degradation under periodic inspections, 160–4
Reliability improvement, 191–212
  knowledge engineering for, 191–212
Repair model, 137–40, 146
Replacement model, 139, 140
Residual life distribution, 160

Safety conditions, color-face graphic display, 204–9
Safety index, damaged redundant structures, 107
Safety margin, damaged redundant structures, 107
Scatter factors
  distribution function of, 29
  known Weibull shape parameter, 27–36
  reliability-based, 25–50

Scatter factors—*contd.*
unknown shape and scale parameters, 36–47
Simulation analysis, reliability of bridge structures, 119–34
S–N curve, 223
Specified failure probability, 126–7, 130, 132
Stochastic crack growth model, 154
Stochastic Finite Element Method, 214, 220
Stochastic model, 137
Stress analysis of notched plate, 58–62, 67–9
Stress intensity factor, 71

Tanker hull, transverse ring, 97–102
Tensor polynomial theory, 55
Three Mile Island, 198, 205–9
Time history response analysis, 220
Transmission line tower, 94–7

Ultimate collapse analysis, frame structures, 85–117
Unstable fracture, 21
Upper bound approach, probability of failure, 18–22

Weibull distribution, 27, 36, 75–7, 159
Weibull shape parameter, 27–36
WELCON, 192